ISSN 0044-2267

ZEITSCHRIFT FÜR ANGEWANDTE MATHEMATIK UND MECHANIK

INGENIEURWISSENSCHAFTLICHE FORSCHUNGSARBEITEN

BAND 61 1981 HEFT 8

AUS DEM INHALT

HAUPTAUFSÄTZE

AKADEMIE-VERLAG · BERLIN

ZAMM Bd. 61 Nr. 8 S. 345—408 Berlin, August 1981

EVP 18,— M 34115

INHALT

Wir bitten, **Manuskriptsendungen** zweifach (Original und eine Kopie, sprachlich einwandfrei, Formeln mit Maschine oder in Druckschrift geschrieben) an folgende Anschrift zu richten:

Zeitschrift für Angewandte Mathematik und Mechanik,
Institut für Mechanik der Akademie der Wissenschaften der DDR
DDR-1199 Berlin, Rudower Chaussee 5.

Zu den Arbeiten, die als Hauptaufsätze bestimmt sind, ist auf gesondertem Blatt eine Zusammenfassung von 5 bis 10 Zeilen in englischer und (möglichst) deutscher und russischer Sprache beizufügen. Ausführliche Hinweise für die Autoren, um deren strikte Berücksichtigung gebeten wird, finden sich im Anschluß an das Inhaltsverzeichnis des Jahrganges **60** (1980).

Die Autoren erhalten von den Hauptaufsätzen 75, von den Kleinen Mitteilungen 25 Sonderdrucke ohne Berechnung, darüber hinaus weitere Sonderdrucke gegen Berechnung.

Der Verlag behält sich für alle Beiträge das Recht der Vervielfältigung und Übersetzung vor.

Bestellungen sind zu richten

- **in der DDR** an den Postzeitungsvertrieb unter Angabe der Kundennummer des Bestellenden oder an den AKADEMIE-VERLAG, DDR-1080 Berlin, Leipziger Straße 3—4
- **im sozialistischen Ausland** an eine Buchhandlung für fremdsprachige Literatur oder an den zuständigen Postzeitungsvertrieb
- **in der BRD und Berlin(West)** an eine Buchhandlung oder an die Auslieferungsstelle KUNST UND WISSEN, Erich Bieber OHG, D-7000 Stuttgart 1, Wilhelmstraße 4—6
- **in den übrigen westeuropäischen Ländern** an eine Buchhandlung oder an die Auslieferungsstelle KUNST UND WISSEN, Erich Bieber GmbH, Ch-8008 Zürich/Schweiz, Dufourstraße 51
- **im übrigen Ausland** an den Internationalen Buch- und Zeitschriftenhandel; den Buchexport, Volkseigener Außenhandelsbetrieb der Deutschen Demokratischen Republik, DDR-7010 Leipzig, Postfach 160; oder an den AKADEMIE-VERLAG, DDR-1080 Berlin, Leipziger Straße 3—4

ZEITSCHRIFT FÜR ANGEWANDTE MATHEMATIK UND MECHANIK

Herausgeber und Chefredakteur: Prof. Dr. Günter Schmidt. Redaktion: Dr. Winfried Heinrich, Dr. Horst Weinert, Dipl.-Math. Friedhild Dudel, Helga Rühl, Institut für Mechanik der Akademie der Wissenschaften der DDR. Verlag: Akademie-Verlag, DDR-1080 Berlin, Leipziger Straße 3—4; Fernruf: 2236221 oder 2236229. Telex-Nr.: 114420; Bank: Staatsbank der DDR, Berlin, Kto.-Nr.: 6836-26-20712. Anschrift der Redaktion: Institut für Mechanik der Akademie der Wissenschaften, DDR-1199 Berlin, Rudower Chaussee 5, Fernruf: 6702841.

Veröffentlicht unter der Lizenznummer 1282 des Presseamtes beim Vorsitzenden des Ministerrates der Deutschen Demokratischen Republik. Gesamtherstellung: VEB Druckerei „Thomas Müntzer", DDR-5820 Bad Langensalza. Erscheinungsweise: Die Zeitschrift für Angewandte Mathematik und Mechanik erscheint monatlich. Die 12 Hefte eines Jahres einschließlich Tagungshefte bilden einen Band. Bezugspreis je Band 360,— M zuzüglich Versandspesen (Preis für die DDR 216,— M). Bezugspreis je Heft 30,— M (Preis für die DDR 18,— M). Bestellnummer dieses Heftes: 1009/61/8.

AN (EDV) 35937

ASMW

ISSN 0044-2267

ZAMM 61, 349 – 358 (1981)

K. Z. Markov

On the Dilatation Theory of Elasticity

In dieser Arbeit wird die lineare Theorie des Festkörpers, für welche das Verschiebungsfeld und die Volumendilatation unabhängig sind, in Einzelheiten erforscht. Die Theorie ist für eine Beschreibung des Elastizitätskörpers, der eine große Zahl Mikroporen enthält, angewandt. Für diesen Fall wurden die in der Theorie vorkommenden Materialkonstanten berechnet.

In this paper a linear theory for a solid, for which the displacement and dilatation (volume change) fields are independent, is investigated in details. The theory is applied to describe the mechanical behaviour of an elastic solid containing a large number of microvoids and the material constants, entering the theory, are calculated for this case.

Работа посвящена детальному изучению линейной теории твердого деформируемого тела, для которого поля перемещений и дилятации (объемного изменения) независимы. Теория применена для описания механического поведения упругого тела содержащего большое количество микропор. Для этого случая подсчитаны материальные постоянные, участвующие в теории.

1. Introduction

In the classical continuum mechanics there is a fundamental assumption that the basic characteristic of the motion is the field $\boldsymbol{u}$ of the body points displacement. This field is basic in the sense that all the rest deformation characteristics can be found provided the field $\boldsymbol{u}$ is known. For example, under the supposition of small strains the simple relations hold true, viz.

$$\boldsymbol{\Phi} = \tfrac{1}{2} \nabla \times \boldsymbol{u}\,, \qquad \Theta = \nabla \cdot \boldsymbol{u} \tag{1.1}$$

which describe the rotation $\boldsymbol{\Phi}$ and dilatation Θ for a small volume of the body. If however the microstructure of the material is taken into account, we should reject the relations (1.1) and consider the fields $\boldsymbol{\Phi}$ and Θ as certain independent characteristics of deformation.

The assumption that the rotation field $\boldsymbol{\Phi}$ does not depend on the displacement field $\boldsymbol{u}$ has been first proposed by Voigt [1]. Such an assumption leads to the well known micropolar theory of elasticity which has been widely investigated during the last two decades (cf., e.g., [2], [3]). To the best of our knowledge, except for author's paper [4], the assumption of the independence between the dilatation and displacement fields has not been closely considered though the corresponding theory it leads to and which could be called dilatation elasticity may be useful when analyzing problems concerning mechanical behaviour and fracture for materials, containing a large number of point defects and/or microvoids.

The presented paper deals with the formalism and possible interpretation for the isotropic linear dilatation elasticity. The basic premises of the theory are given in Sec. 2; the presentation there is closely connected to [4]. In Sec. 3 the plane problem for the dilatation-elastic body is discussed. As a simple example the stress and strain fields around an edge dislocation in the body are constructed there. For the sake of completeness, the basic equations for a micropolar body with independent dilatation of the body points are derived in Sec. 4. In Sec. 5 an attempt is made to interpret the dilatation elasticity as a theory for an elastic solid containing a large number of microvoids. It appears then that the dilatation elasticity is a theory of first-grade type which is able to account for the nonlocal interactions between voids solely by means of the first gradient of the void ratio. Within the frame of such an interpretation a try is performed in Sec. 6 to calculate explicitly the five material constants that appear in the dilatation elasticity. The method to calculate these constants is based on certain requirements of equivalence for two simple solutions for microporous and dilatation bodies respectively.

2. Basic Equations for the Dilatation Elasticity

In the classical continuum mechanics the interaction between two parts of a body is described by a field of forces, acting on the surface that divides these two parts and characterized by the Cauchy stress tensor $\boldsymbol{T}_\sigma$. The mean triaxial stress p in a point of the body is easily expressed through the trace of the stress tensor: $p = \frac{1}{3} \operatorname{tr} \boldsymbol{T}_\sigma$.

In the dilatation elasticity the field of body points dilatation is supposed not to depend on the displacement field. This supposition makes the kinematics of the body more complicated compared to the classical elasticity, because we have to describe the motion of the body under consideration by means of a vector field $\boldsymbol{u}$ which gives the displacements for the body points and of a scalar field Θ which gives the dilatation for the points. Any complication of kinematics always yields complication of dynamics as well. In conformity to the case we are discussing this means, first of all, that the mean triaxial stress s becomes independent of the stress tensor, so that some additional mean triaxial stress $\sigma = s - p \neq 0$ should appear as a result of the presence of the "free dilatation" $\Theta^c = \Theta - \nabla \cdot \boldsymbol{u} \neq 0$.

In order to describe the additional mean triaxial stress $\sigma = s - p$ (for brevity, we shall call it dilatation stress in the sequel), we assume that a vector field $\boldsymbol{\lambda}$ exists which possesses the following characteristic property: the integral $\int_S \boldsymbol{\lambda} \cdot \mathrm{d}\boldsymbol{S}$ over each closed surface (S) gives the dilatation stress acting on the body volume inside (S) on the part of the body volume outside (S). The field $\boldsymbol{\lambda}$ is naturally to be named the vector field of dilatation stress.

As it is clear from its definition, the field $\boldsymbol{\lambda}$ is introduced by analogy with the CAUCHY stress tensor. In this way the "dilatation forces", which appear within the body due to the presence of free dilatation, are tacitly supposed to be contact — to have zero-radius of action, i.e. to possess the same nature as the CAUCHY stresses.

The equation of the equilibrium at a point of the body under consideration is the same as that in the classical continuum mechanics, viz.

$$\nabla \cdot \boldsymbol{T}_\sigma + \varrho \boldsymbol{f} = 0; \tag{2.1}$$

the case of statics will be merely of interest in what follows. In (2.1) $\boldsymbol{f}$ is the density of the body forces. Note that the dilatation stress does not generate body moments. That is why the stress tensor $\boldsymbol{T}_\sigma$ remains symmetric.

By the very definition of the vector field $\boldsymbol{\lambda}$ of dilatation stresses, the relation $\int_V (s - p)\,\mathrm{d}V = \int_S \lambda_n\,\mathrm{d}S$ holds for each part (V) of the body; here (S) is the surface of (V), $\lambda_n = \boldsymbol{\lambda} \cdot \boldsymbol{n}$ and $\boldsymbol{n}$ is the outward normal unit vector to (S). The last relation yields that

$$\nabla \cdot \boldsymbol{\lambda} + p - s = 0 \tag{2.2}$$

which is the second equilibrium equation for the body under consideration. Equation (2.2) expresses the balance of the mean triaxial stress at a point of the body.

Suppose there exists a potential energy of the reversible deformation $W = W(\boldsymbol{u}, \Theta)$. Making use of the virtual work principle and of equations (2.1) and (2.2), we have

$$\begin{aligned}
&\int_V \delta W\,\mathrm{d}V = \int_S \sigma_n \cdot \delta\boldsymbol{u}\,\mathrm{d}S + \int_S \lambda_n \delta\Theta\,\mathrm{d}S + \int_V \varrho\boldsymbol{f} \cdot \delta\boldsymbol{u}\,\mathrm{d}V = \\
&= \int_V (\nabla \cdot \boldsymbol{T}_\sigma + \varrho\boldsymbol{f}) \cdot \delta\boldsymbol{u}\,\mathrm{d}V + \int_V \boldsymbol{T}_\sigma\colon \delta(\boldsymbol{u}\nabla)\,\mathrm{d}V + \int_V \nabla \cdot \boldsymbol{\lambda}\delta\Theta\,\mathrm{d}V + \int_V \boldsymbol{\lambda} \cdot \delta\nabla\Theta\,\mathrm{d}V = \\
&= \int_V (\boldsymbol{T}_\sigma\colon \delta(\boldsymbol{u}\nabla) + (s - p)\,\delta\Theta + \boldsymbol{\lambda} \cdot \delta\nabla\Theta)\,\mathrm{d}V\,.
\end{aligned}$$

Since the stress tensor $\boldsymbol{T}_\sigma$ is symmetric, the latter relation leads to the formula

$$\delta W = \boldsymbol{T}_\sigma\colon \delta\boldsymbol{T}_e + (s - p)\,\delta\Theta + \boldsymbol{\lambda} \cdot \delta\nabla\Theta \tag{2.3}$$

for the variation δW at a point; here, $\boldsymbol{T}_e = \frac{1}{2}(\nabla\boldsymbol{u} + \boldsymbol{u}\nabla)$ is the small strain tensor, the colon means contraction with respect to two pairs of indices, e.g. $\boldsymbol{T}_\sigma\colon \delta\boldsymbol{T}_e = \sigma_{ij}\delta e^{ij}$. The potential W is thus a function of the arguments $\boldsymbol{T}_e$, Θ and $\nabla\Theta$

$$W = W(\boldsymbol{T}_e, \Theta, \nabla\Theta)\,. \tag{2.4}$$

Having varied (2.4) and compared the result with (2.3), we receive the general form of the stress-strain relation for the reversible small-strain deformation for the body under consideration, namely

$$\boldsymbol{T}_\sigma = \frac{\partial W}{\partial \boldsymbol{T}_e}, \qquad s - p = \frac{\partial W}{\partial \Theta}, \qquad \boldsymbol{\lambda} = \frac{\partial W}{\partial \nabla\Theta}\,. \tag{2.5}$$

The set of equations (2.1), (2.2), (2.4) and (2.5) specifies a theory which we call dilatation elasticity.

It is to note that the dilatation elasticity could be found as a particular case of the so-called micromorphic theory, proposed by ERINGEN et al. [5], under the assumption that the microdistortion tensor is spherical. However, the considerations made above are simpler than those used in [5], because the usage of microelements and averaging over macroelements were avoided here.

Hereafter we shall conisder merely the case of an isotropic solid. In this case the potential W has the form

$$W = \tfrac{1}{2}(\lambda + \eta)\,e^2 + \mu\,\mathrm{tr}\,\boldsymbol{T}_e^2 - \eta\Theta e + \tfrac{1}{2}\delta\Theta^2 + \tfrac{1}{2}\chi|\nabla\Theta|^2\,, \tag{2.6}$$

where $e = \mathrm{tr}\,\boldsymbol{T}_e$ is the volume change due to the elastic deformation. Thus, there are five material constants λ, μ, η, δ and χ in the linear isotropic dilatation elasticity. Taking into account (2.5) and (2.6) we get now the following linear stress-strain relations

$$\boldsymbol{T}_\sigma = (\lambda e - \eta\Theta^c)\,\boldsymbol{I} + 2\mu\boldsymbol{T}_e\,, \qquad \boldsymbol{\lambda} = \chi\nabla\Theta\,, \qquad s - p = \delta\Theta^c + (\delta - \eta)\,e\,, \tag{2.7}$$

where $\boldsymbol{I}$ is the unit 2nd order tensor and $\Theta^c = \Theta - e$ is, to remind, the free dilatation.

Note that the condition $W > 0$ of positiveness for the potential W is equivalent to the following inequalities

$$\mu,\ k + \eta,\ \delta,\ \chi\ (k + \eta)\,\delta - \eta^2 > 0\,, \qquad (k = \lambda + \tfrac{2}{3}\mu)\,. \tag{2.8}$$

Inserting (2.7) into (2.1) and (2.2), we find the system of the LAMÉ type for the isotropic dilatation elasticity, viz.

$$\begin{aligned}
&\mu\,\Delta\boldsymbol{u} + (\lambda + \mu)\,\nabla\nabla \cdot \boldsymbol{u} - \eta\nabla\Theta^c + \varrho\boldsymbol{f} = 0\,, \\
&\chi\,\Delta\Theta - \delta\Theta + \eta\nabla \cdot \boldsymbol{u} = 0\,.
\end{aligned} \tag{2.9}$$

Here $\Delta = \nabla \cdot \nabla$ is the Laplacian.

For the sake of completeness we shall derive as well the equations of the dilatation elasticity in respect of the stress characteristics $\boldsymbol{T}_\sigma$ and $\boldsymbol{\lambda}$.

Following IGNACZAK [6], we first apply to $(2.9)_1$ the operator def, i.e. the operator of symmetrized gradient of a vector field: $\boldsymbol{v} \to \mathrm{def}\,\boldsymbol{v} = \frac{1}{2}(\nabla\boldsymbol{v} + \boldsymbol{v}\nabla)$, then

$$\mu\,\Delta\boldsymbol{T}_e + (\lambda + \mu)\,\nabla\nabla e - \eta\nabla\nabla\Theta^c + \varrho\,\mathrm{def}\,\boldsymbol{f} = 0\,.$$

Having solved $(2.7)_1$ with respect to the strain tensor, we get

$$2\mu \boldsymbol{T}_e = \boldsymbol{T}_\sigma - \frac{1}{1+\nu}(3\nu p - \eta(1-2\nu)\,\Theta^c)\,\boldsymbol{I}\,, \qquad e = \frac{p + \eta\Theta^c}{k}, \tag{2.10}$$

$\nu = \dfrac{\lambda}{2(\lambda+\mu)}$, which we insert into the previous relation to obtain that

$$\Delta \boldsymbol{T}_\sigma - \frac{3\nu}{1+\nu}\boldsymbol{I}\Delta p + \frac{3}{1+\nu}\nabla\nabla p + \eta\frac{1-2\nu}{1+\nu}(\boldsymbol{I}\,\Delta\Theta^c + \nabla\nabla\Theta^c) + \varrho\,\operatorname{def}\boldsymbol{f} = 0\,. \tag{2.11}$$

This equation can be simplified if we apply the operator div to $(2.9)_1$: $(\lambda + 2\mu)\,\Delta e - \eta\,\Delta\Theta^c + \varrho\nabla\cdot\boldsymbol{f} = 0$ and use then $(2.10)^2$ in order to exclude e. Then we can get rid of Δp in (2.11). Eventually,

$$\Delta \boldsymbol{T}_\sigma + \frac{3}{1+\nu}\nabla\nabla p + \frac{\nu}{1-\nu}\varrho\boldsymbol{I}\nabla\cdot\boldsymbol{f} + 2\varrho\,\operatorname{def}\boldsymbol{f} + \eta\frac{1-2\nu}{1-\nu}\boldsymbol{I}\Delta\Theta^c + \eta\frac{1-2\nu}{1+\nu}\nabla\nabla\Theta^c = 0\,. \tag{2.12}$$

It is noteworthy that (2.12) reduces to the well-known BELTRAMI equation from the linear elasticity provided the field Θ^c of free dilatation is linear function of Cartesian co-ordinates.

The equation for the dilatation stress vector $\boldsymbol{\lambda}$ could readily be found having applied the operator ∇ to $(2.9)_2$. Eventually,

$$\Delta\boldsymbol{\lambda} - q_0^2\boldsymbol{\lambda} + \frac{\eta}{k+\eta}\nabla p = 0\,, \tag{2.13}$$

where $q_0^2 = \dfrac{(k+\eta)\,\delta - \eta^2}{(k+\eta)\,\chi}$. Note that always $q_0^2 > 0$, as it follows from (2.8).

The equations (2.12) and (2.13) form a system of nine equations which specifies the stress characteristics $\boldsymbol{T}_\sigma$ and $\boldsymbol{\lambda}$ for an isotropic dilatation elastic solid.

To bring this Section to a conclusion we note that the GREEN tensor and general solution for the system (2.9) could be constructed in a more or less standard manner, having decomposed the displacement field as a sum of two potentials: $\boldsymbol{u} = \boldsymbol{b} + \nabla\psi$. Here, we give the final results only. The first one concerns the displacement $\boldsymbol{U}_e(\boldsymbol{x}-\boldsymbol{x}')$ and dilatation $\Theta_e(\boldsymbol{x}-\boldsymbol{x}')$ fields, generated by unit force acting along the direction $\boldsymbol{e}$, $|\boldsymbol{e}| = 1$, and concentrated at the point $\boldsymbol{x}'$. They are of the form

$$\boldsymbol{U}_e = \boldsymbol{U}\cdot\boldsymbol{e}\,, \qquad \boldsymbol{U} = \frac{1}{4\pi\mu}\left(\frac{1}{R}\boldsymbol{I} - \frac{q_2}{2q_1}\nabla\nabla R + \frac{\mu\chi\eta^2}{mq_1^2}\nabla\nabla h(R/m)\right),$$

$$\Theta_e = \boldsymbol{D}\cdot\boldsymbol{e}\,, \qquad \boldsymbol{D} = \frac{\eta}{4\pi q_1 m}\nabla h(R/m); \qquad h(s) = \frac{1}{s}(1-e^{-s})\,, \qquad R = |\boldsymbol{x}-\boldsymbol{x}'|\,, \tag{2.14}$$

with the constants $m^2 = \chi\dfrac{\lambda+2\mu+\eta}{q_1}$, $q_1 = (\lambda+2\mu+\eta)\,\delta - \eta^2$, $q_2 = (\lambda+\mu+\eta)\,\delta - \eta^2$. The second result refers to the general solution of the PAPKOVICH-NEUBER type for the system (2.9). It has the form

$$\boldsymbol{u} = \boldsymbol{b} + \nabla\left(\frac{\eta\chi}{q_1}\Theta - \frac{q_2}{2q_1^2}\boldsymbol{r}\cdot\boldsymbol{b} + b_0\right), \qquad \Theta = \Theta_0 + \frac{\eta\mu}{q_1}\nabla\cdot\boldsymbol{b}\,, \tag{2.15}$$

where $\boldsymbol{r} = \boldsymbol{x}$, $\boldsymbol{b}$ and b_0 are arbitrary harmonic functions and Θ_0 is an arbitrary solution of the equation $m^2\,\Delta\Theta - \Theta = 0$. For details, concerning (2.14) and (2.15), cf. [4].

3. Plane Problem for the Dilatation Elasticity

In the case of plane deformation the stress and strain fields are functions of two co-ordinates; for simplicity, we choose them as the Cartesian co-ordinates x and y. The first equilibrium equation (2.1) is the same as that in the classical elasticity. That is why it can be satisfied by introducing the AIRY function $\varphi = \varphi(x, y)$ [7], so that

$$\sigma_x = \frac{\partial^2\varphi}{\partial y^2}\,, \qquad \tau_{xy} = -\frac{\partial^2\varphi}{\partial x\,\partial y}\,, \qquad \sigma_y = \frac{\partial^2\varphi}{\partial x^2}\,. \tag{3.1}$$

We shall use in what follows the AIRY function φ and the free dilatation Θ^c as basic functions (potentials) for the plane problem in discussion. Body forces are supposed to vanish, $\boldsymbol{f} = 0$.

It is to note first the relation

$$(\lambda + 2\mu)\,\Delta e - \eta\,\Delta\Theta^c = 0 \tag{3.2}$$

which is an obvious consequence of $(2.9)_1$. Making use of $(2.9)_2$ and (3.2), we get the following equation

$$m^2\,\Delta\Delta\Theta^c - \Delta\Theta^c = 0 \tag{3.3}$$

for the field of free dilatation.

On the other hand, we have

$$e = \nabla \cdot u = \frac{p + \eta \Theta^c}{\lambda + \mu}, \qquad p = \frac{1}{2} \operatorname{tr} \boldsymbol{T}_\sigma = \frac{1}{2} \Delta\varphi, \tag{3.4}$$

as it follows from (2.7) and (3.1).

Equations (3.2) to (3.4) yield now that the function

$$\psi = \varphi + \frac{2\mu\eta}{\lambda + 2\mu} m^2 \Theta^c \tag{3.5}$$

is biharmonic, i.e. $\Delta\Delta\psi = 0$.

It is worth mentioning another relation between the potentials φ and Θ^c. It follows from (3.2), (3.4) and $(2.9)_2$ and has the form

$$\frac{\chi(\lambda + 2\mu + \eta)}{\lambda + 2\mu} \Delta\Theta^c - \frac{q_2}{\lambda + \mu} \Theta^c + \frac{\eta - \delta}{2(\lambda + \mu)} \Delta\varphi = 0\,. \tag{3.6}$$

We are intended now to derive formulae of the Love type which would give us the displacements in a dilatation elastic body through the potentials φ and Θ^c.

With that aim in view, we first rewrite $(2.7)_1$ in the form

$$2\mu \boldsymbol{T}_e = \boldsymbol{T}_\sigma - 2\nu p \boldsymbol{I} + (1 - 2\nu)\, \eta \Theta^c \boldsymbol{I}\,. \tag{3.7}$$

Let u_x and u_y be the displacements along the axes x and y respectively. Since $e_x = \frac{\partial u_x}{\partial x}$ and $e_y = \frac{\partial u_y}{\partial y}$, we have the relations

$$\begin{aligned} 2\mu \frac{\partial u_x}{\partial x} &= -\frac{\partial^2 \varphi}{\partial x^2} + (1 - \nu)\, \Delta\varphi + (1 - 2\nu)\, \eta\Theta^c\,, \\ 2\mu \frac{\partial u_y}{\partial x} &= -\frac{\partial^2 \varphi}{\partial y^2} + (1 - \nu)\, \Delta\varphi + (1 - 2\nu)\, \eta\Theta^c\,, \end{aligned} \tag{3.8}$$

reached by making use of (3.1), (3.4) and (3.7). Next we replace the function φ in (3.8) through the function ψ according to (3.5), then

$$\begin{aligned} 2\mu \frac{\partial u_x}{\partial x} &= -\frac{\partial^2 \varphi}{\partial x^2} + (1 - \nu)\, \Delta\psi - (1 - 2\nu)\, \eta(m^2 \Delta\Theta^c - \Theta^c)\,, \\ 2\mu \frac{\partial u_y}{\partial y} &= -\frac{\partial^2 \varphi}{\partial y^2} + (1 - \nu)\, \Delta\psi - (1 - 2\nu)\, \eta(m^2 \Delta\Theta^c - \Theta^c)\,. \end{aligned} \tag{3.9}$$

But both functions

$$P = \Delta\psi\,, \qquad R = m^2 \Delta\Theta^c - \Theta^c \tag{3.10}$$

are harmonic, according to (3.5) and (3.3) respectively. We can thus introduce the analytical functions

$$F(z) = P + iQ\,, \qquad G(z) = R + iS \tag{3.11}$$

of the complex argument $z = x + iy$, so that Q and S should be the harmonic functions conjugated with the functions P and R respectively.

Consider the functions

$$f(z) = \tfrac{1}{4} \int F(w)\, dw = f_1 + if_2\,, \qquad g(z) = \tfrac{1}{4} \int G(w)\, dw = g_1 + ig_2\,, \tag{3.12}$$

so that

$$P = \Delta\psi = 4\frac{\partial f_1}{\partial x} = 4\frac{\partial f_2}{\partial y}, \qquad R = m^2 \Delta\Theta^c - \Theta^c = 4\frac{\partial g_1}{\partial x} = 4\frac{\partial g_2}{\partial y}.$$

Having used these two relations in (3.9), we find eventually that

$$\begin{aligned} 2\mu u_x &= -\frac{\partial \varphi}{\partial x} + 4(1 - \nu) f_1 - 4(1 - 2\nu)\, \eta g_1 + \text{const}\,, \\ 2\mu u_y &= -\frac{\partial \varphi}{\partial y} + 4(1 - \nu) f_2 - 4(1 - 2\nu)\, \eta g_2 + \text{const}\,. \end{aligned} \tag{3.13}$$

This is the desired generalization of the Love formulae [8] for the case of a dilatation elastic solid undergoing plane deformation.

As a simple example of application of Love's type relations (3.13), we shall consider the displacement field generated by an edge dislocation in an unbounded dilatation elastic body.

Let ϱ, α be the polar co-ordinates, so that $\alpha = 0$ be the slip direction and $\varrho = 0$ — the centre of the dislocation.

We first choose the biharmonic function ψ from (3.5) to be the same as that in the case of elastic material [9], namely,

$$\psi = \varphi + \frac{2\mu\eta}{\lambda + 2\mu} m^2 \Theta^c = a_1 \varrho \ln \varrho \sin \alpha , \tag{3.14}$$

and the solution of (3.3) for the free dilatation as follows

$$\Theta^c = \left(\frac{a_2}{\varrho} + a_3 K_1(\varrho/m)\right) \sin \alpha + a_4 . \tag{3.15}$$

Here a_j, $j = 1$ to 4, are some constants and K_1 stands for the McDONALD function of the first order.

The analytical functions (3.11), which correspond to the potentials (3.14) and (3.15), are very simple, viz.

$$F(z) = \frac{2a_1 i}{z}, \qquad G(z) = -\frac{i a_2}{z}. \tag{3.16}$$

Making use of (3.12) to (3.16), we find after simple calculations that

$$\begin{aligned} 2\mu u_x &= -\frac{\partial \varphi}{\partial x} - (2a_1(1-\nu) + \eta(1-2\nu)\, a_2)\, \alpha + \text{const} , \\ 2\mu u_y &= -\frac{\partial \varphi}{\partial y} - (2a_1(1-\nu) + \eta(1-2\nu)\, a_2) \ln \varrho + \text{const} . \end{aligned} \tag{3.17$_1$}$$

The relations (3.17) give the displacement field generated by an edge dislocation in a dilatation elastic solid. The constants a_1 and a_2 should be specified through the two facts we know: first, the BURGERS vector b is given and second, the functions φ and Θ^c have to satisfy (3.6). Eventually, we get

$$a_1 = -\frac{q_2}{2q_2(1-\nu) + \eta(\eta-\delta)(1-2\nu)} \frac{\mu b}{\pi}, \qquad a_2 = \frac{\eta - \delta}{q_2} a_1 \tag{3.17$_2$}$$

fo that the known result for the displacement field around an edge dislocation in elastic body [9] leads from (3.17) sor either $\eta = 0$ or $\eta = \delta$.

It is to note that the constants a_3 and a_4 remain undetermined in (3.17). The reason is clear: we merely have the condition of the jump for u_x, i.e. the value of the BURGERS vector; nothing is required for the free dilatation field Θ^c. For instance, at least formally, we could require the field Θ^c to possess the least possible singularity at the dislocation centre $\varrho = 0$. This requirement would yield that $ma_3 = -a_2$. In this case the displacement field would correspond to a special distribution of "dilatation sources" along the line $\alpha = \frac{1}{2}\pi$ and it would have logarithmical behaviour in the vicinity of the dislocation centre. At last, the constant a_4 could be readily specified provided the average value $\langle\Theta^c\rangle$ of the free dilatation in the body is prescribed. Then $a_4 = \langle\Theta^c\rangle$, as it follows from (3.15).

4. Micropolar-Dilatation Body

For the sake of completeness, we shall briefly consider in this section a model for a deformable body which could be called micropolar-dilatation. For such a model the rotation $\boldsymbol{\Phi}$ and dilatation Θ of the body points are both supposed not to depend on the displacement field $\boldsymbol{u}$. Let us remind that the independence of the rotation and displacement fields leads to the appearance, first, of a skew-symmetric part for the stress tensor and, second, of the couple-stress tensor $\boldsymbol{T}_\mu$ [2], [3]. The equation of the equilibrium for the linear momentum has its usual form (2.1). The equilibrium for the moment of linear momentum is expressed by the equation

$$\nabla \cdot \boldsymbol{T}_\mu - \boldsymbol{E} : \boldsymbol{T}_\sigma + \boldsymbol{m} = 0 \tag{4.1}$$

[2], [3], where $\boldsymbol{m}$ is the volume density of the body moments and $\boldsymbol{E}$ is the RICCI alternator.

Using the virtual work principle we get in just the same way as in section 2 that

$$\delta W = \boldsymbol{T}_\sigma : \delta \boldsymbol{T}_\gamma + \boldsymbol{T}_\mu : \delta \boldsymbol{T}_\varkappa + (s - p)\, \delta\Theta + \boldsymbol{\lambda} \cdot \delta \nabla \Theta ,$$

where

$$\boldsymbol{T}_\gamma = \nabla \boldsymbol{u} + \boldsymbol{E} \cdot \boldsymbol{\Phi} , \qquad \boldsymbol{T}_\varkappa = \nabla \boldsymbol{\Phi}$$

are the strain tensors for a micropolar body [2], [3]. That is why $W = W(\boldsymbol{T}_\gamma, \boldsymbol{T}_\varkappa, \Theta, \nabla\Theta)$ and, for small strain,

$$\boldsymbol{T}_\sigma = \frac{\partial W}{\partial \boldsymbol{T}_\gamma}, \qquad \boldsymbol{T}_\mu = \frac{\partial W}{\partial \boldsymbol{T}_\varkappa}, \qquad s - p = \frac{\partial W}{\partial \Theta}, \qquad \boldsymbol{\lambda} = \frac{\partial W}{\partial \nabla \Theta} \tag{4.2}$$

with the notations of section 2: to remind, s is the full mean triaxial stress at a point, $p = \frac{1}{3} \operatorname{tr} \boldsymbol{T}_\sigma$ and $\boldsymbol{\lambda}$ stands for the vector of dilatation stresses.

The set of equations (2.1), (2.2), (4.1) and (4.2) specifies a theory which we call micropolar-dilatation elasticity.

It is to note that the micropolar-dilatation elasticity could also be found as a particular case of the above mentioned micromorphic theory [5], under the assumption that the microdistortion tensor is a linear combination of a spherical and a skew-symmetrical tensors. Some other theories which could be obtained from the micromorphic theory under special assumptions about the microdistortion tensor have been discussed in [13].

For a centrosymmetric body and small strains the potential W has the following quadratic form

$$W = \tfrac{1}{2}(\lambda + \eta)(\mathrm{tr}\, \boldsymbol{T}_\gamma)^2 + \mu\, \mathrm{tr}\,(\boldsymbol{T}_\gamma^s \cdot \boldsymbol{T}_\gamma^s) + \alpha\, \mathrm{tr}\,(\boldsymbol{T}_\gamma^a \cdot \boldsymbol{T}_\gamma^a) + \tfrac{1}{2}\beta(\mathrm{tr}\, \boldsymbol{T}_\varkappa)^2 +$$
$$+ \gamma\, \mathrm{tr}\,(\boldsymbol{T}_\varkappa^s \cdot \boldsymbol{T}_\varkappa^s) + \varepsilon\, \mathrm{tr}\,(\boldsymbol{T}_\varkappa^a \cdot \boldsymbol{T}_\varkappa^a) - \eta\Theta\, \mathrm{tr}\, \boldsymbol{T}_\gamma + \tfrac{1}{2}\delta\Theta^2 + \tfrac{1}{2}\chi|\nabla\Theta|^2 . \tag{4.3}$$

The indices "s" and "a" denote here the symmetric and skew-symmetric parts, respectively, for the corresponding tensors. Note that the tensor $\boldsymbol{T}_\varkappa$ changes its sign under inversion. Thus the mutual invariants of $\boldsymbol{T}_\varkappa$ with the rest of the strain characteristics, i.e. with $\boldsymbol{T}_\gamma$, Θ and $\nabla\Theta$ which are invariant with respect to inversion, do not enter (4.3).

Having inserted (4.3) into (4.2), we receive the following stress-strain relations for the body under consideration

$$\begin{aligned}
&\boldsymbol{T}_\sigma = (\lambda e - \eta\Theta^c)\, \boldsymbol{I} + 2\mu \boldsymbol{T}_\gamma^s + 2\alpha \boldsymbol{T}_\gamma^a ,\\
&\boldsymbol{T}_\mu = \beta \boldsymbol{I}\, \mathrm{tr}\, \boldsymbol{T}_\varkappa + 2\gamma \boldsymbol{T}_\varkappa^s + 2\varepsilon \boldsymbol{T}_\varkappa^a ,\\
&\boldsymbol{\lambda} = \chi\nabla\Theta , \qquad s - p = \delta\Theta^c + (\delta - \eta)\, e \qquad (e = \mathrm{tr}\, \boldsymbol{T}_\gamma = \nabla \cdot \boldsymbol{u}) .
\end{aligned} \tag{4.4}$$

The equations of equilibrium (2.1), (2.2) and (4.1) together with the stress-strain relations (4.4) yield now the equations of the LAMÉ type for the centrosymmetric micropolar-dilatation solid

$$\begin{aligned}
&(\mu + \alpha)\, \Delta \boldsymbol{u} + (\lambda + \mu - \alpha)\, \nabla\nabla \cdot \boldsymbol{u} + 2\alpha\nabla \times \boldsymbol{\Phi} - \eta\nabla\Theta^c = 0 ,\\
&(\gamma + \varepsilon)\, \Delta\boldsymbol{\Phi} + (\beta + \gamma - \varepsilon)\, \nabla\nabla \cdot \boldsymbol{\Phi} - 4\alpha\boldsymbol{\Phi} + 2\alpha\nabla \times \boldsymbol{u} = 0 ,\\
&\chi\Delta\Theta - \delta\Theta + \eta\nabla \cdot \boldsymbol{u} = 0 ,
\end{aligned} \tag{4.5}$$

supposing that the body forces and moments vanish.

It is important to point out that the dilatation field is presented in $(4.5)_1$ by its gradient and the rotation field — by its rotor. Hence, the dilatation and rotation fields do not interact one with other. This remark immediately leads to the general solution of the system (4.5), viz.

$$\begin{aligned}
&\boldsymbol{u} = \boldsymbol{b} + \nabla\left(\frac{\eta\chi}{q_1}\Theta - \frac{q_2}{2q_1^2}\boldsymbol{r} \cdot \boldsymbol{b} + \frac{2\alpha}{\mu} l_1^2 \nabla \cdot \boldsymbol{\psi} + \chi\hat{o}\right) + \frac{2\alpha}{\mu}\boldsymbol{\psi} ,\\
&\boldsymbol{\Phi} = \nabla \times \left(\frac{1}{2}\boldsymbol{b} + \frac{\alpha + \mu}{\mu}\boldsymbol{\psi}\right) + \nabla\psi_0 , \qquad \Theta = \Theta_0 + \frac{\eta\mu}{q_1}\nabla \cdot \boldsymbol{b} ,
\end{aligned} \tag{4.6}$$

having added the NEUBER representation [10] for the displacement and rotation fields in a micropolar body to the general representation (2.15) for a dilatation elastic body. The functions $\boldsymbol{\psi}$ and ψ_0, entering (4.6), solve the HELMHOLTZ equations [10]: $l_1^2\, \Delta\boldsymbol{\psi} - \boldsymbol{\psi} = 0$, $l_2^2 \Delta\psi_0 - \psi_0 = 0$, where $l_1^2 = \dfrac{(\mu + \alpha)\,(\gamma + \varepsilon)}{4\alpha\mu}$, $l_2^2 = \dfrac{\beta + 2\gamma}{4\alpha}$.

Note that the construction of the GREEN tensor for the system (4.5) is not difficult due to the just mentioned independence of the rotation and dilatation fields (cf. [4] for details).

5. An Interpretation of the Dilatation Elasticity

Consider a solid containing a great number of micropores. We suppose the micropores are spherical and homogeneously distributed so that the solid should be macroscopically homogeneous and isotropic.

Let $\Theta^c(\boldsymbol{x})$ be the porosity (void ratio) at a point $\boldsymbol{x}$ of the body. As usual, point means here the representative elementary volume which size has to be, on the one hand, much bigger than the average size of the voids and, on the other hand, much smaller than the size of the specimen. As we neglected from the very beginning the details concerning shape and distribution of the voids, the field $\Theta^c(\boldsymbol{x})$ presents an exhaustive description for damaging of the solid. We assume too that the solid is microdamaged, i.e. $\Theta^c \ll 1$. Thus, the porous sintered metals whose void ratio could attain 0,30 to 0,40 will not be under discussion in what follows. Note that in the latter case the influence of the void ratio becomes apparent, first of all, through an essential decrease of the overall elastic moduli (see, e.g., [11], [12]). In the former case of a microdamaged solid, the decrease of the overall moduli due to the void ratio is negligible. The main effect caused by the void ratio consists then in appearance of microstresses which, even when having small magnitude, should be considered to play a leading role in the mechanical behaviour of a microdamaged specimen [14].

It is important to note that the case of a microdamaged material, which we are going to discuss, is typical for solids at state, generally speaking, not very close to rupture. It is well known that defects of void type appear at the very beginning of the deformation process. For metals, this is a result of dislocation junctions or of accumulation of discrepancy at transition of slip planes through low-angle boundaries (cf. [9]). In view of the specific nature of the metal state, it is no wonder that the magnitude of the void ratio for metals is rather small; as a rule, it is not bigger than 10^{-3}. An enumeration of the various mechanisms which could generate microporosity ("loosening") of metals as well as some experimental facts concerning the magnitude of their microporosity are given in [15]. However, despite the microporosity is a trifle, there are weighty arguments that its magnitude governs fracture, in the sense that the fracture of a specimen occurs when its microporosity attains a certain critical value [15], [16], [17].

The foregoing remarks justify any reasonable attempt to derive a continuum theory for a microporous solid. Here we shall try to describe the mechanical behaviour of such a solid by using the dilatation elastic model, discussed in section 2. It is clear that the pivotal point of such an endeavour should consist in taking the free dilatation to coincide with the microporosity (we have already taken the liberty of denoting these two quantities by one and the same symbol Θ^c).

With that purpose in view, we first put an obvious requirement: at the absence of both free dilatation and body forces, the equations (2.9) of the dilatation elasticity must be consequences of the LAMÉ equations of the linear elasticity. It is readily seen that this requirement is equivalent to the relation

$$\eta = \delta\,, \tag{5.1}$$

so that the system (2.9) could be rewritten in the form

$$\mu\,\Delta u + (\lambda + \mu)\,\nabla\nabla\cdot u - \eta\nabla\Theta^c = 0\,, \tag{5.2}$$
$$\chi'\Delta\Theta^c - \eta\Theta^c = 0\,. \qquad \left(\chi' = \chi\left(1 + \frac{\eta}{\lambda + 2\mu}\right)\right).$$

It is worth mentioning that the same equality (5.1) yields from the ESHELBY theorem [18] which states that in the expression for the elastic energy of a solid the term that gives the interaction between the external and internal stresses vanishes. In fact, if a homogeneous distribution of free dilatation (i.e. microporosity) be considered, then, according to (2.6),

$$W = \tfrac{1}{2}(k + \eta)\,e^2 - \eta\Theta e + \tfrac{1}{2}\delta\Theta^2 = \tfrac{1}{2}ke^2 + \tfrac{1}{2}(\delta - \eta)\,e^2 + \underline{(\delta - \eta)\,e\Theta^c} + \tfrac{1}{2}\delta(\Theta^c)^2\,.$$

The underlined term gives the interaction between the external and internal stresses. Its vanishing, required by the theorem of ESHELBY, leads once again to equality (5.1).

As the void ratio $\Theta^c \ll 1$, by supposition, the constants λ and μ, entering (2.6), are just the LAMÉ moduli for the solid part of the microporous material under consideration. The main problem that now appears consists in calculation of the material constants η and χ through what we know about the solid, i.e. through the LAMÉ moduli of its solid part and the void ratio Θ^c.

Before we go at calculating the material constants η and χ, it is necessary that some general remarks, concerning the dilatation elasticity as a theory of a microdamaged solid, should be made.

Consider a nonhomogeneous distribution of micropores in a solid. It is known that the interaction between the micropores (which are simulated by simplest singular solutions of elasticity [7]) is then nonlocal, at the expence of the effects of image [18]. Thus, the density of the elastic energy at a point x of the microporous solid depends not only on the strain tensor and magnitude of void ratio in $\boldsymbol{x}$ but also on the values $\Theta^c(\cdot)$ of the void ratio all over the body, i.e. W should be a functional of the form

$$W = W(\boldsymbol{T}_e, \Theta^c(\cdot))\,. \tag{5.3}$$

To the best of our knowledge no attempts have been made to calculate explicitly the functional (5.3) for bounded bodies. Note that the functional (5.3) has been calculated merely for several types of crystals in which nonlocality of void interactions results from anisotropy [19].

Comparison of the functional (5.3) for the elastic energy of a microdamaged solid with the analogous expression (2.4) for a dilatation elastic body sheds light upon the limits of applicability of the dilatation elasticity as a feasible continuum model for the microporous solid. As it is clear from (2.4), the dilatation elasticity could be only a first approximation for the full nonlocal theory of such a solid, in which the nonlocality makes appearance through the first gradient $\nabla\Theta^c$. In this way, the dilatation elasticity is related to the so-called strain-gradient theories in which the nonlocal continuum interactions are roughly accounted for by introducing the gradients of the strain tensor into the HOOKE law [20]. That is why the limits of applicability for the dilatation elasticity should be similar to those for the strain-gradient theories, namely, the dilatation elasticity should be expected to describe, at least in a qualitative way, the mechanical behaviour or a microporous solid provided the magnitude of the microporosity gradient $\nabla\Theta^c$ is small enough. In view of this fact, when estimating the dilatation constants χ and η in the next section, we shall restrict our analysis to some qualitative speculations (keeping in mind that strict considerations within the frames of a theory, which is approximate by its very origin, are often pure fiction).

6. Calculation of the Material Constants η and χ

In this section we consider a method for explicit calculation of the constants η and χ in the dilatation elastic model of a microporous solid. The method is based on certain equivalence requirements for two pairs of simple solutions found for microporous and dilatation elastic bodies respectively.

6.1. Microporous Sphere under Hydrostatic Pressure

Consider a sphere $(V_R) = \{\boldsymbol{x} \mid r < R\}$, $r = |\boldsymbol{x}|$, with a radius R, undergoing an external hydrostatic pressure P, so that

$$\sigma_r|_{r=R} = -P\,. \tag{6.1}$$

We suppose the average porosity Θ_0 of the sphere to be known, i.e.

$$\langle\Theta^c\rangle = \frac{1}{V_R}\int\limits_{V_R} \Theta^c(\boldsymbol{x})\,\mathrm{d}V = \Theta_0 \tag{6.2}$$

and $\Theta_0 \ll 1$. We shall determine the displacement field into the sphere V_R making use of the equations (5.2) of the dilatation elasticity, under the conditions (6.1) and (6.2).

As the problem we deal with is centrosymmetric, we have

$$u = \nabla f , \qquad f = f(r) , \tag{6.3}$$

which allows to rewrite (5.2) in the form

$$(\lambda + 2\mu)\, \Delta f - \eta \Theta^c = a_1 , \qquad \chi' \,\Delta \Theta^c - \eta \Theta^c = 0 . \tag{6.4}$$

Eliminating Θ^c from (6.4), we get

$$f = \frac{\chi'}{\lambda + 2\mu} \Theta^c + \frac{a_1}{6(\lambda + 2\mu)} r^2 . \tag{6.5}$$

The solution of the HELMHOLTZ equation $(6.4)_2$ bounded at $r = 0$ in our case has the form

$$\Theta^c(r) = a_2 \frac{\sinh r/m'}{r/m'} , \qquad m' = \sqrt{\chi'/\eta} . \tag{6.6}$$

The constants a_1 and a_2 in (6.5) and (6.6) have to be chosen so that the conditions (6.1) and (6.2) are satisfied, which gives

$$\begin{aligned} a_1 &= \frac{\lambda + 2\mu}{k}\left(-P + \frac{4}{3}\frac{\mu\eta}{\lambda + 2\mu}\Theta_0\right), \\ a_2 &= \frac{\Theta_0}{3D} , \qquad D = \frac{1}{d^3}(d \cosh d - \sinh d) , \qquad d = R/m' . \end{aligned} \tag{6.7}$$

It is clear from (6.6) and $(6.7)_2$ that the usage of the equations (5.2) of the dilatation elasticity leads to a non-homogeneous distribution for the free dilatation Θ^c within the sphere under hydrostatic pressure. And what is more, the magnitude of the free dilatation near the surface $r = R$ is always bigger than that at the centre $r = 0$.

Keeping in mind (6.3) and (6.5) to (6.7), we find after some algebra the following expression

$$u_r|_{r=R} = f(R) = -\frac{P}{3k} R + \frac{1}{3} \Theta_0 \frac{\eta}{\lambda + 2\mu}\left(1 + \frac{4\mu}{3k}\right) R \tag{6.8}$$

for the displacement of the points situated on the surface $r = R$.

The first term on the right side of (6.8) presents the displacement for the surface points of an elastic sphere, undergoing hydrostatic pressure [7]. The second term should be considered as a result of the expansion of the sphere due to the existence of free dilatation. On the other hand, a simplest analysis shows that the expansion of a sphere with average porosity Θ_0 is $\frac{1}{3}\Theta_0 R$ which, equated to the second term entering the right side of (6.8), yields

$$\eta = k . \tag{6.9}$$

In other words, let us consider two identical spheres, one made of porous elastic material with average void ratio Θ_0, the other — of dilatation elastic material with average value of the free dilatation $\langle \Theta^c \rangle$, so that $\langle \Theta^c \rangle = \Theta_0$. Then the relation (6.9) ensures the "equivalence" of those two spheres in the following sense. We cannot tell which one is made of elastic and which one of dilatation elastic material, using merely measuring of their surface displacement under hydrostatic pressure — because this displacement would be one and the same for both spheres provided (6.9) holds true.

As such an equivalence represents a natural requirement as far as we want to simulate the mechanical behaviour of a microporous elastic solid through the dilatation elastic model, we shall consider (6.9) as the desired calculation for the first of the two dilatation constants we are seeking for.

It is to note that the supposition of microporosity $\Theta_0 \ll 1$ for the elastic sphere was used above only to provide the bulk modulus k in (6.9) to coincide with that of the solid part of the sphere. If the void ratio Θ_0 is not small, the foregoing considerations remain valid, with the only change that the bulk modulus $k = k(\Theta_0)$ for the porous solid should enter (6.9). Explicit expressions for function $k(\Theta_0)$ could be taken from various theories, predicting the overall moduli for porous elastic solids (cf., e.g., [11], [12]).

It should be pointed out that, as a matter of fact, the same equality (6.9) has been obtained by ESHELBY, who has derived equation $(5.2)_1$ and also noted the analogy between body expansion generated by temperature increasing and that caused by damaging [18].

In order to specify the last unknown dilatation constant χ, we should examine another problem, concerning a microporous elastic solid and its simulation by means of the dilatation elastic model.

6.2. Spherical Hole in Unbounded Microporous Solid

Consider an unbounded dilatation elastic material containing a spherical hole with a radius R. The material undergoes an internal pressure P acting on the surface of the hole, so that the same boundary condition (6.1) holds. We then look for the solution of the system (5.2).

In the case in discussion, the free dilatation field Θ^c has the form

$$\Theta^c = a_3 \frac{e^{-r/m'}}{r/m'} , \qquad r > R; \qquad m' = \sqrt{\chi'/\eta} , \tag{6.10}$$

— cf. $(6.4)_2$. For the potential f, according to (6.5) and (6.10), we get

$$f = \frac{\chi'}{\lambda + 2\mu} \Theta^c - \frac{a_4}{r}. \tag{6.11}$$

Keeping in mind (6.10), (6.11) and $(2.7)_1$, we obtain the radial stress σ_r to be

$$\sigma_r = -\frac{4\mu\eta}{\lambda + 2\mu} \frac{D_1(x)}{x} a_3 - \frac{2\mu}{r^3} a_4, \qquad r > R, \tag{6.12}$$

where

$$D_1(x) = \frac{e^{-x}}{x}\left(1 + \frac{1}{x}\right), \qquad x = r/m'.$$

In order to satisfy the boundary condition (6.1), the unknown constants a_3 and a_4 should comply with the relation

$$\frac{4\mu k}{\lambda + 2\mu} \frac{D_1(d)}{d} a_3 + \frac{2\mu}{R^3} a_4 = P, \qquad d = R/m'. \tag{6.13}$$

As our basic idea in this section consists in simulating of the elastic porous material by means of a dilatation elastic one, we are bound to examine now the same unbounded body containing the spherical hole, but made of elastic material whose void ratio changes r in accordance with (6.10).

As it can be seen from (6.10), the magnitude of the void ratio Θ^c is not small in the vicinity of the hole. This fact makes us take into account the decrease of the elastic moduli due to porosity. Here, in order to simplify the calculations in what follows, we shall apply the relations

$$\mu^* = \mu(1 - 2\Theta^c), \qquad k^* = k(1 - 2\Theta^c) \tag{6.14}$$

for the shear and bulk moduli respectively for a porous solid with void ratio Θ^c, having assumed for simplicity that the POISSON ratio of the solid part $\nu = 0.2$ [12], [21].

However, the solution of the centrosymmetric problem considered for a body, whose elastic moduli vary along r in accordance with (6.14) and (6.10), cannot be explicitly done. That is the reason to make one more simplification supposing that the void ratio $\Theta^c(r)$ is replaced by a step-constant function $\tilde{\Theta}^c(r)$ such that

$$\tilde{\Theta}^c(r) = \begin{cases} \frac{1}{2} a_3 D_0, & R < r < L, \\ 0, & r > L, \end{cases} \tag{6.15}$$

$D_0 = e^{-d}/d$, with the value $\frac{1}{2} a_3 D_0$ which is a half of the maximal one for the function $\Theta^c(r)$, given by (6.10). The parameter L is to be found from the obvious condition

$$\int_{r>R} \Theta^c(r)\, dV = \int_{r>R} \tilde{\Theta}^c(r)\, dV,$$

which specifies L to be

$$(L/m')^3 = d^3 + 6d(1 + d). \tag{6.16}$$

Consider now a solid whose elastic constants μ^* and k^* are defined by means of (6.14) and (6.16). The centrosymmetric problem we deal with, namely, the expansion of a spherical hole undergoing internal pressure in an unbounded matrix, can be readily solved for such a partially homogeneous elastic solid. The solution looks as follows

$$\sigma_r = P_0 \frac{L^3}{r^3}, \qquad r > L; \qquad u_r = A_1 r + \frac{A_2}{r^2}, \qquad R < r < L, \tag{6.17}$$

where

$$P_0 = \frac{P}{1 + 6g\Theta^*}, \qquad g = \frac{1 + d}{d^2}, \qquad \Theta^* = \frac{1}{2} a_3 D_0,$$

$$A_1 = \frac{PR^3 - P_0 L^3}{3k^*(L^3 - R^3)}, \qquad A_2 = \frac{R^3 L^3 (P - P_0)}{4\mu^*(L^3 - R^3)}.$$

We have two solutions, the first — (6.10) to (6.12) and the second — (6.17), for one and the same problem obtained within the frames of the dilatation and classical elasticity respectively. Since the void ratio of the elastic material has been chosen above in accordance with the free dilatation field (6.10), it is natural certain conditions of identity for the two solutions to be required. The simplest conditions of such a type are, for one thing, the equality of the displacements on the surface of the hole $r = R$ and, for another thing, one and the same asymptotic behaviour for the stress fields at infinity. These two requirements lead to the relations

$$-2\mu a_4 = P_0 R^3, \qquad A_1 R + \frac{A_2}{R^2} = \frac{\chi'}{\lambda + 2\mu} \frac{d\Theta^c}{dr}(R) + \frac{a_4}{R^2}$$

which, together with (6.13) and (6.16), form a set of equations for determining the three unknown parameters a_3, a_4 and $d = R/m'$. After some algebra, we obtain that $d = R/m' = 2/\sqrt{s}$, $s = P/\mu$, having assumed that $s \ll 1$ which sounds fully reasonable for real hydrostatic pressure. As $m' = \sqrt{\chi'/k}$ (cf. (6.9) and (6.10)), the latter result yields that

$$\chi' = \frac{1}{4} kR^2 \frac{P}{\mu}. \tag{6.18}$$

So, the value of the dilatation constant χ' appears to depend on the radius R, entering the subsidiary problem we have just examined. A gross estimation for this value could be obtained by supposition that the radius R coincides with a certain characteristic length for the microporous solid. But the basic such a length for this kind of solid is the average radius a of the voids. Taking $R = a$, we get from (6.18) the final result

$$\chi' = \chi\left(1 + \frac{k}{\lambda + 2\mu}\right) = \frac{1}{4}\frac{k}{\mu}|I_1|\, a^2, \quad \text{i.e.} \quad \chi = \frac{2}{9}|I_1|\, a^2, \tag{6.19}$$

where the hydrostatic pressure P from (6.18) is replaced by the first invariant for the stress tensor: $I_1 = \operatorname{tr} \boldsymbol{T}_\sigma$ and (5.2) is taken into account as well as our assumption that the POISSON ratio for the solid part $\nu = 0{,}2$.

The relations (5.1), (6.9) and (6.19) complete the calculation of the material constants, entering the dilatation elastic model for a microporous elastic solid.

7. Concluding Remarks

In conclusion, it is necessary that a very essential feature of the equality (6.19) should be underlined, namely, the fact that the dilatation constant χ has appeared to depend on the stress tensor. This circumstance makes the basic system (5.2) nonlinear, i.e. the dilatation elastic model for a microporous solid becomes nonlinear. In general, the ratio $|P|/\mu < 0{,}01$ because we have to expect that the pressure applied to the body is not greater than the yield stress, which leads to the possibility of rejecting the effect of microporosity. The latter effect, however, should be of great value in the regions in which stress concentration appears, e.g. near the tip of a crack or core of dislocation. Thus, if our intention consists in analyzing the void concentration in the vicinity of a crack or dislocation, then the strongly nonlinear system (5.2) should be applied, with the values of the dilatation parameters taken from (5.1), (6.9) and (6.19). This represents an extremly difficult nonlinear problem within which, moreover, plastic yielding should be also taken into consideration.

Consequently, the linear dilatation elasticity seems to be a very gross approximation as far as our purpose is to simulate the behaviour of a microdamaged solid. Such a solid should be better simulated through a nonlinear dilatation model for which the material constant $\chi = \chi(I_1)$ is a function of the mean triaxial stress. The equation (6.19) gives a possible form for such a function.

References

1 VOIGT, W., Theoretische Studien über die Elastizitätsverhältnisse der Kristalle, Abh. Ges. Wiss. Göttingen, **34**, 1887.
2 NOWACKI, W., Theory of micropolar elasticity, Springer, 1970.
3 ERINGEN, A. C., Theory of micropolar elasticity, in: Fracture, ed. by H. LIEBOWITZ, vol. 2, Acad. Press, N.Y.-London 1968.
4 MARKOV, K. Z., On the theory of elasticity of bodies with independent dilatation of particles [in Russian], Theoretical and Applied Mechanics (Sofia), **5** (1974), No 1, pp. 93—99.
5 ERINGEN, A. C.; SUHUBI, E. S., Nonlinear theory of micro-elastic solids, Int. J. Engng Sci., **2** (1964), pp. 189—203 and 389—404.
6 IGNACZAK, J., Direct determination of stress equations of motion in elasticity, Arch. Mech. Stosow., **11** (1959), pp. 672—678.
7 SOKOLNIKOFF, I. S., Mathematical theory of elasticity, Mc Graw Hill, New York 1956.
8 LOVE, A. E. H., The mathematical theory of elasticity, Cambridge University Press, Cambridge 1927.
9 HIRTH, J.; LOTHE, J., Theory of dislocations, Mc Graw-Hill, New York 1968.
10 NEUBER, H., On the general solution of linear elastic problems in isotropic and anisotropic COSSERAT continua, Proc. 11th International Congress on Applied Mechanics, Munich 1964, Springer-Verlag, 1966, pp. 153—158.
11 MACKENZIE, J. K., The elastic constants of a solid containing spherical holes, Proc. Phys. Soc., **B 63** (1950), No 1, pp. 2—11.
12 SKOROCHOD, V. V., Calculation of the elastic isotropic moduli for particulate solid mixtures [in Russian], Poroshkovaja metallurgia, 1961, No 1, pp. 50—55.
13 MARKOV, K. Z., Fundamental solutions in the micromorphic elasticity, in: Theoretical and Applied Mechanics (Proc. 2nd National Congress on Mechanics, Varna, Bulgaria, 1973), Sofia 1975, vol. 1, pp. 184—190.
14 NOVOJILOV, V. V.; KADASHEVICH, YU., Theory of plasticity taking into account the residual microstresses [in Russian], Prikladnaja matematika i mehanika (PMM), **22** (1958), No 1.
15 KOLMOGOROV, V. L., ed., Plasticity and fracture [in Russian], Metalurgia, Moscow 1977.
16 MCCLINTOCK, F. A.; ARGON, A., Deformation and fracture of materials, Mc Graw-Hill, New York 1968.
17 ARUTYUNYAN, R. A.; MARKOV, K. Z., Loozening and fracture of plastic solids, Letters Appl. Engng Sci., **5** (1977), pp. 187—193.
18 ESHELBY, J. D., Continuum theory of defects ,in: Solid State Physics, vol. **3**, New York 1956, p. 79.
19 HACHATURYAN, A. G., Theory of phase transformations and structure of solid solutions [in Russian], Nauka, Moscow 1974.
20 KUNIN, I. A., Theory of elastic media with microstructure (nonlocal elasticity) [in Russian], Nauka, Moscow 1975.
21 BUDIANSKY, B., On the elastic moduli of some heterogeneous materials, J. Mech. Phys. Solids, **13** (1965), pp. 223—227.

Eingereicht am 15. 10. 1980

Anschrift: Associate Professor KONSTANTIN MARKOV CSc, Centre for Mathematics and Mechanics, Bulgarian Academy of Sciences, Sofia 1090, P. O. Box 373, Bulgaria

ZAMM 61, 359–364 (1981)

A. Chakrabarti / K. Manivachakan

On the Transform Method of Solution of an External Crack Problem

The breakdown of the usual method of Fourier transforms in the problem of an external line crack in a thin infinite elastic plate is discovered and the correct solution of this problem is derived using the concept of a generalised Fourier transform of a type discussed first by Golecki [1] in connection with Flamant's problem.

Es wird das Versagen der gewöhnlichen Fouriertransformation im Falle eines äußeren Linienrisses bei einer unendlichen elastischen Platte nachgewiesen und eine korrekte Lösung dieses Problems hergeleitet, indem das Konzept einer verallgemeinerten Fouriertransformation benutzt wird, welches zuerst von Golecki [1] in Verbindung mit dem Flamantschen Problem diskutiert wurde.

Доказана неприменимость обыкновенного преобразования Фурье в случае внешней линии трещины в бесконечной гибкой пластине. Корректное решение этой задачи построено с помощью обобщённого преобразования Фурье, которое обсуждалось впервые Голецким [1] в связи с проблемой Фламанта.

1. Introduction

The use of the Fourier transform (FT) method employed in solving two-dimensional isothermal boundary value problems of elastostatics has been made simple by Sneddon [2] in deriving the following representation of the displacement components $u_x(x, y)$ and $u_y(x, y)$:

$$u_x = \frac{\partial \chi}{\partial x} + \frac{\partial \varphi}{\partial x} + (\beta^2 - 1)\, y\, \frac{\partial^2 \varphi}{\partial y\, \partial x}\,, \qquad u_y = \frac{\partial \chi}{\partial y} - \beta^2 \frac{\partial \varphi}{\partial y} + (\beta^2 - 1)\, y\, \frac{\partial^2 \varphi}{\partial y^2} \tag{1), (2}$$

where

$$\nabla^2 \varphi = 0 = \nabla^2 \chi; \qquad \beta^2 = \frac{2(1 - \sigma)}{1 - 2\sigma}; \tag{3}$$

σ being the Poisson ratio of the homogeneous isotropic elastic material. The other representations of the displacement field are in terms of Airy stress function (see pp. 204—205 of [3]) or in terms of its FT, G, the latter being given by (see pp. 402—404 of [4]):

$$u_x = \frac{1 + \sigma}{2\pi E} \int\limits_{-\infty}^{\infty} \left[(1 - \sigma) \frac{d^2 G}{dy^2} + \sigma \xi^2 G \right] i\, e^{-i\xi x} \frac{d\xi}{\xi}, \tag{4}$$

$$u_y = \frac{1 + \sigma}{2\pi E} \int\limits_{-\infty}^{\infty} \left[(1 - \sigma) \frac{d^3 G}{dy^3} + (2 + \sigma)\, \xi^2 \frac{dG}{dy} \right] e^{-i\xi x} \frac{d\xi}{\xi^2} \tag{5}$$

where E is the Young's modulus and

$$G(\xi, y) = (A + By)\, e^{-|\xi| y} + (C + Dy)\, e^{|\xi| y}\,.$$

Here A, B, C, D are in general functions of ξ and are determined from the boundary conditions of the problem. The representations (4)—(5) for u_x and u_y are obtained under the assumption that the components of stress and displacement have FT (w.r.t. x). The representations (1)—(2) or (4)—(5) have been the starting point of the method of FT (see [5] and [6]) in a number of investigations on two-dimensional crack problems in classical theory of elasticity.

In the present paper, we demonstrate that the usual FT method breaks down in the case of the problem of an external line crack in a thin infinite elastic plate. We then introduce the concept of a generalised Fourier transform (GFT) on the lines of Golecki [1] (see also [7], [8]) and show that the difficulties encountered in the usual transform method applied to such problems can be overcome by using our GFT. We also observe that we do not have to formulate these boundary value problems through dual integral equations as is customary in the usual transform method. In fact, such a formulation is shown to create difficulties in the simple problem considered in this paper and perhaps it may also do so in more general problems of this category. Such difficulties arise, perhaps, as Erdogan [9] felt in a different context, "due to the improper use of integral transforms for the functions which may not be transformable in the conventional sense" (see also [10]). Golecki's paper [1] in which he explains how the formal application of FT leads to solutions expressed by divergent integrals, by considering the example of Flamant's problem and by showing that the displacement component normal to the boundary and the Airy stress function for this problem do not have FT in the usual sense, supports the contention of Erdogan [9]. Similar divergent integrals also occur in the FT method of solution of the crack problem discussed in this paper.

After briefly demonstrating the breakdown of the usual FT method in Section 2, we introduce a GFT in Section 3. We then apply this GFT, in Section 4, where we obtain useful expressions for the derivatives of the displacement components on the boundary of a half-plane when the boundary is loaded by prescribed stresses. The external crack problem (ECP) is then considered in Section 5 where this problem is reduced in a natural way to a singular integral equation of the first kind with a CAUCHY-type kernel. The solution of this integral equation is written down using standard results. We observe that the final results obtained are in agreement with the solution of the problem via complex variable method. However, the solution derived through the usual transform method differs from the above and is not the correct solution for the problem.

2. The usual transform method for ECP

The boundary value problem associated with an external line crack in a thin infinite elastic plate was tackled by LOWENGRUB [11] through FT method and a more general problem was discussed earlier through complex variable method by ERDOGAN [12], [13]. The boundary conditions, in this case, after reducing the problem to that for the half-plane ($y > 0$) are:

$$\sigma_{xy}(x, 0) = 0\,, \qquad -\infty < x < \infty\,, \tag{6}$$

$$\sigma_{yy}(x, 0) = -p(x)\,, \qquad |x| \geqq 1\,, \qquad u_y(x, 0) = 0\,, \qquad |x| < 1\,, \tag{7), (8}$$

$p(x)$ being a prescribed function. The infinity conditions assumed are:

(i) The stress-state at infinity is zero, i.e., the three stress components all vanish as $x^2 + y^2 \to \infty$.
(ii) The rotation at infinity is zero.
(iii) The plane is loaded at infinity and the resultant of the external forces at infinity is prescribed, viz.,

$$\int_{-\infty}^{\infty} \sigma_{yy}(x, 0)\, \mathrm{d}x = Y_\infty \tag{9}$$

where Y_∞ is specified.

Using the representations (1)—(2) or (4)—(5) (see pp. 405—406 of [4]), LOWENGRUB [11] assumed the displacement field in the following form:

$$u_x = \frac{1}{2(1-\sigma)} \int_0^\infty \xi^{-1}\, \mathrm{e}^{-\xi|y|}(1 - 2\sigma - \xi|y|)\, [\psi_a(\xi) \cos \xi x - \psi_s(\xi) \sin \xi x]\, \mathrm{d}\xi\,, \tag{10}$$

$$u_y = \frac{1}{2(1-\sigma)} \int_0^\infty \xi^{-1}\, \mathrm{e}^{-\xi|y|}(2 - 2\sigma + \xi|y|)\, [\psi_s(\xi) \cos \xi x + \psi_a(\xi) \sin \xi x]\, \mathrm{d}\xi \tag{11}$$

where the two unknown functions ψ_s and ψ_a are made to satisfy the following set of dual integral equations:

$$\int_0^\infty \xi^{-1}\psi_s(\xi) \cos \xi x\, \mathrm{d}\xi = 0\ (0 \leqq x < 1); \qquad \int_0^\infty \psi_s(\xi) \cos \xi x\, \mathrm{d}\xi = \frac{2(1-\sigma^2)}{E} p_s(x) \qquad (x > 1)\,, \tag{12}$$

$$\int_0^\infty \xi^{-1}\psi_a(\xi) \sin \xi x\, \mathrm{d}\xi = 0 \qquad (0 \leqq x < 1); \qquad \int_0^\infty \psi_a(\xi) \sin \xi x\, \mathrm{d}\xi = \frac{2(1-\sigma^2)}{E} p_a(x) \qquad (x > 1) \tag{13}$$

in order to satisfy the boundary conditions (6)—(8). Here,

$$p_s(x) = \tfrac{1}{2}\,[p(x) + p(-x)]\,; \qquad p_a(x) = \tfrac{1}{2}\,[p(x) - p(-x)]\,. \tag{14}$$

The sets of dual integral equations (12) and (13) are solved in [11] by LOWENGRUB. The solution of (12) is given there as:

$$\psi_s(\xi) = \frac{4(1-\sigma^2)}{E\pi} \left\{ \int_1^\infty p_s(x) \cos \xi x\, \mathrm{d}x - [1 - J_0(\xi)] \int_1^\infty p_s(x)\, \mathrm{d}x - \xi \int_0^1 J_1(\xi t)\, \mathrm{d}t \int_1^\infty \frac{x p_s(x)\, \mathrm{d}x}{(x^2 - t^2)^{1/2}} \right\}. \tag{15}$$

In the case of symmetrical loading

$$p_s(x) = P\delta(c - x); \qquad p_a(x) = 0 \qquad (c > 1)\,, \qquad |x| > 1\,,$$

discussed in detail in [11], we find that $\psi_s(0) = \dfrac{4(1-\sigma^2)\, P}{E\pi} \neq 0$. This clearly shows the divergence of the integral on the right of (11) which represents $u_y(x, y)$ as also that on the left of the first integral equation of (12). We thus observe that the formulation of the problem through the dual integral equations is not alright. As stated earlier, the integral expressions assumed for u_x und u_y in (10) and (11) are obtained under the assumption that u_x and u_y have FT. It is well known that the existence of FT for a function imposes a certain restriction on the behaviour of the function at infinity. In the absence of any stipulation regarding the behaviour of u_x and u_y at infinity, it is not advisable to assume the existence of their FT. On the other hand, as the stress components are assumed to vanish as $|x| \to \infty$, it may be assumed that these components have FT (w.r.t. x) and this in turn implies the existence of FT of the first order derivatives of u_x and u_y. This provides the motivation for generalising the FT as described in the following section.

3. First order generalised Fourier transform (First oder GFT)

In this section, we introduce the concept of a GFT, $F(\xi, y)$, of a function $f(x, y)$ of two variables, through the usual FT of the first order derivatives of f, instead of through the FT of its second order derivatives, as defined by GOLECKI [1], which (the latter) we shall call as second order GFT.

Definition: Let $f(x, y)$ be a function which is such that its partial derivatives $\partial f/\partial x$ and $\partial f/\partial y$ are absolutely integrable and absolutely continuous in $-\infty < x < \infty$. Then we call the function $F(\xi, y)$, defined below, the First order GFT of $f(x, y)$ w.r.t. x.

$$-i\xi F = \int_{-\infty}^{\infty} \frac{\partial f}{\partial x} \mathrm{e}^{i\xi x} \, \mathrm{d}x \,. \tag{16}$$

By FOURIER inversion formula, we then have

$$\frac{\partial f}{\partial x} = \frac{-i}{2\pi} \int_{-\infty}^{\infty} \xi F \, \mathrm{e}^{-i\xi x} \, \mathrm{d}\xi \,. \tag{17}$$

If we make a further assumption that the second order derivatives of f are absolutely continuous and absolutely integrable w.r.t. x in $-\infty < x < \infty$, we deduce from (16), that

$$\frac{\partial F}{\partial y} = \int_{-\infty}^{\infty} \frac{\partial f}{\partial y} \mathrm{e}^{i\xi x} \, \mathrm{d}x \tag{18}$$

and then by FOURIER inversion,

$$\frac{\partial f}{\partial y} = \frac{1}{2\pi} \int_{-\infty}^{\infty} \frac{\partial F}{\partial y} \mathrm{e}^{-i\xi x} \, \mathrm{d}\xi \,. \tag{19}$$

The two relations (17) and (19) suffice to determine the function $f(x, y)$ by integrating the relation

$$\mathrm{d}f = \frac{\partial f}{\partial x} \mathrm{d}x + \frac{\partial f}{\partial y} \mathrm{d}y \tag{20}$$

except for an arbitrary constant, from the knowledge of the GFT, $F(\xi, y)$, of f, which then serves as an inversion formula for the GFT.

Remark: In case, when f is also absolutely integrable, we find by integrating (16) by parts, that

$$F(\xi, y) = \hat{f}(\xi, y)$$

where $\hat{f}(\xi, y)$ is the ordinary FT of f w.r.t. x defined by

$$\hat{f}(\xi, y) = \int_{-\infty}^{\infty} f(x, y) \, \mathrm{e}^{i\xi x} \, \mathrm{d}x \,. \tag{21}$$

This means that whenever a function has FT in the usual sense, it has also the GFT which is identical with its usual FT but not vice-versa.

4. Displacement-derivatives on the boundary of a half-plane

In this section, we obtain useful expressions for the derivatives $\frac{\partial u_x}{\partial x}$, $\frac{\partial u_x}{\partial y}$, $\frac{\partial u_y}{\partial x}$ and $\frac{\partial u_y}{\partial y}$ (by using the representations (1)—(2)) on the boundary $y = 0$ of the half-plane $y > 0$ when the boundary is loaded as:

$$\sigma_{xy}(x, 0) = -f_1(x) \,, \qquad -\infty < x < \infty; \qquad \sigma_{yy}(x, 0) = -f_2(x) \,, \qquad -\infty < x < \infty \tag{22), (23}$$

where f_1 and f_2 are assumed to be absolutely integrable in $-\infty < x < \infty$ and the stress components are assumed to vanish as $|x|$, $y \to \infty$.

Assume that FT exist for the stress components σ_{xx}, σ_{yy} and σ_{xy}. Denote the First order GFT of u_x, u_y and the Second order GFT of φ, χ (in the representations (1)—(2)) by U_x, U_y, Φ and X respectively. The LAPLACE's equation

$$\frac{\partial^2 \varphi}{\partial x^2} + \frac{\partial^2 \varphi}{\partial y^2} = 0 \tag{24}$$

satisfied by $\varphi(x, y)$, by an obvious procedure, transforms into

$$\left(\frac{\mathrm{d}^2}{\mathrm{d}y^2} - \xi^2\right) \Phi = 0 \,. \tag{25}$$

We thus observe that even if φ fails to possess its ordinary FT $\hat{\varphi}$, the partial differential equation (24) transforms into the ordinary differential equation (25), as desired by any transform method to be applicable in problems associated with LAPLACE's equation, if the GFT, Φ, exists. Solving (25) we get

$$\Phi(\xi, y) = A(\xi)\, e^{-|\xi| y} + A_1(\xi)\, e^{|\xi| y} \tag{26}$$

and similarly

$$X(\xi, y) = B(\xi)\, e^{-|\xi| y} + B_1(\xi)\, e^{|\xi| y}\,. \tag{27}$$

Again transforming equation (2) to obtain $U_y(\xi, y)$ and then applying (17) to U_y, we obtain

$$\frac{\partial u_y}{\partial x} = \frac{-i}{2\pi} \int\limits_{-\infty}^{\infty} \xi|\xi| \left[e^{-|\xi| y} \{(\beta^2 + (\beta^2 - 1)\, y\, |\xi|)\, A - B\} + e^{|\xi| y} \{(-\beta^2 + (\beta^2 - 1)\, y|\xi|)\, A_1 + B_1\}\right] e^{-i\xi x} d\xi\,. \tag{28}$$

Similar expressions can be derived for the other displacement components and hence for the stress components, by using the stress-strain relations. The condition that the stress components vanish as $y \to \infty$ demands $A_1 = B_1 = 0$. Further the conditions (22) and (23) in their transformed form yield

$$A(\xi) = \frac{(1 - 2\sigma)(1 + \sigma)}{E\xi} \left[\frac{i\hat{f}_1}{|\xi|} + \frac{\hat{f}_2}{\xi}\right]; \qquad B(\xi) = \frac{i(1 + \sigma)}{E\xi|\xi|} \hat{f}_1(\xi)\,. \tag{29}$$

We substitute these values in (28) and write $f_1(x)$ and $f_2(x)$ in the form $f(x) = f_s(x) + f_a(x)$ as in (14). We then interchange the order of integration in (28), observing that for $y > 0$, the integrals are uniformly convergent and use the values of the relevant integrals as evaluated in [14] for $y > 0$ (see also [10]). In the expression so obtained for $\frac{\partial u_y}{\partial x}$, we proceed to the limit as $y \to 0$ and make use of the limiting value of the potential integral (see [10])

$$\operatorname*{Lt}_{y \to 0} \int\limits_a^b \frac{y f(t)\, dt}{y^2 + (t - x)^2}\, (a < x < b) = \pi f(x)\,.$$

We then obtain

$$\frac{E\pi}{2(1 + \sigma)} \left.\frac{\partial u_y}{\partial x}\right|_{y \to 0^+} = (1 - \sigma) \int\limits_0^{\infty} \frac{2x f_{2s}(t) + 2t f_{2a}(t)}{t^2 - x^2}\, dt + \frac{(1 - 2\sigma)\,\pi}{2} f_1(x)\,. \tag{30}$$

By a similar procedure, we get the following:

$$\frac{E\pi}{2(1 + \sigma)} \left.\frac{\partial u_x}{\partial x}\right|_{y \to 0^+} = (1 - \sigma) \int\limits_0^{\infty} \frac{2x f_{1s}(t) + 2t f_{1a}(t)}{t^2 - x^2}\, dt - \frac{(1 - 2\sigma)\,\pi}{2} f_2(x)\,, \tag{31}$$

$$\frac{E\pi}{2(1 + \sigma)} \left.\frac{\partial u_x}{\partial y}\right|_{y \to 0^+} = -(1 - \sigma) \int\limits_0^{\infty} \frac{2x f_{2s}(t) + 2t f_{2a}(t)}{t^2 - x^2}\, dt - \frac{(3 - 2\sigma)\,\pi}{2} f_1(x)\,, \tag{32}$$

$$\frac{E\pi}{2(1 + \sigma)} \left.\frac{\partial u_y}{\partial y}\right|_{y \to 0^+} = -\sigma \int\limits_0^{\infty} \frac{2x f_{1s}(t) + 2t f_{1a}(t)}{t^2 - x^2}\, dt - \frac{(1 - 2\sigma)\,\pi}{2} f_2(x)\,. \tag{33}$$

As a check, one can now verify (22) and (23) by using (30)—(33) and the stress-strain relations.

5. Formulation and solution of ECP

For the ECP discussed in Section 2, we have

$$f_1(x) = 0 \quad (-\infty < x < \infty); \qquad f_2(x) = p(x) \quad (|x| \geqq 1)\,. \tag{34}$$

Setting

$$f_2(x) = r(x)\,, \quad \text{for} \quad |x| < 1 \tag{35}$$

the boundary conditions (7) and (8), together with the equation (30) imply that the components $r_s(x)$ and $r_a(x)$ of $r(x)$ must satisfy the following singular integral equations:

$$\int\limits_0^1 \frac{R_s(T)\, dT}{T - X} = -\int\limits_1^{\infty} \frac{2(2X - 1)\, p_s(t)\, dt}{t^2 - (2X - 1)^2}\,, \qquad 0 < X < 1\,, \tag{36}$$

$$\int\limits_0^1 \frac{R_a(T)\, dT}{T - X} = -\int\limits_1^{\infty} \frac{2t p_a(t)\, dt}{t^2 - (2X - 1)^2}\,, \qquad 0 < X < 1 \tag{37}$$

where $r(x) = r(2X - 1) = R(X)$.

Equations (36) and (37) are of the form:

$$\int_0^1 \frac{L(v)\, dv}{v-u} = g(u)\,, \qquad 0 < u < 1 \tag{38}$$

whose general solution is given by (see [15])

$$L(v) = \frac{C}{\pi[v(1-v)]^{1/2}} - \frac{1}{\pi^2}\frac{1}{[v(1-v)]^{1/2}} \int_0^1 \frac{[u(1-u)]^{1/2} f(u)\, du}{u-v} \tag{39}$$

where $C = \int_0^1 L(v)\, dv$.

The solutions of equations (36) and (37), for a given $p(x)$, can be obtained by using (39), which then solves the ECP completely, the constant C being determined by the requirement (9).

In the particular cases

$$\text{(i)}\quad p_s(x) = P\,\delta(C-x); \qquad p_a(x) = 0 \qquad (C > 1)\,, \qquad |x| > 1\,, \tag{40}$$

$$\text{(ii)}\quad p_s(x) = \frac{P}{2}\,\delta(C-x) = p_a(x) \qquad (C > 1)\,, \qquad |x| > 1 \tag{41}$$

where P is a constant, which are discussed by Lowengrub [11], we find, from our above results, that:

$$\sigma_{yy}^{(i)}(x, 0) = \frac{-Y_\infty^{(i)}}{\pi(1-x^2)^{1/2}} + \frac{2PC(C^2-1)^{1/2}}{\pi(1-x^2)^{1/2}\,(C^2-x^2)}\,, \qquad -1 < x < 1\,, \tag{42}$$

$$u_y^{(i)}(x, 0) = (1-\sigma)\left[-Y_\infty \ln|x + (x^2-1)^{1/2}| + P \ln\left|\frac{x(C^2-1)^{1/2} - C(x^2-1)^{1/2}}{x(C^2-1)^{1/2} + C(x^2-1)^{1/2}}\right|\right] \quad \text{for} \quad x > 1\,, \tag{43}$$

$$K^{(i)} = \frac{-Y_\infty}{\pi} + \frac{2PC}{\pi(C^2-1)^{1/2}}\,, \tag{44}$$

$$\sigma_{yy}^{(ii)}(x, 0) = \frac{-Y_\infty}{\pi(1-x^2)^{1/2}} + \frac{P(C^2 - 1^{1/2})}{\pi(1-x^2)^{1/2}\,(C-x)}\,, \qquad -1 < x < 1\,, \tag{45}$$

$$u_y^{(ii)}(x, 0) = (1-\sigma)\left[-Y_\infty \ln|x + (x^2-1)^{1/2}| + \frac{P}{2} \ln\left|\frac{1 + Cx - [(C^2-1)\,(x^2-1)]^{1/2}}{1 + Cx + [(C^2-1)\,(x^2-1)]^{1/2}}\right|\right] \quad \text{for} \quad x > 1\,, \tag{46}$$

$$K^{(ii)} = \frac{-Y_\infty}{\pi} + \frac{P}{\pi}\left(\frac{C+1}{C-1}\right)^{1/2} \tag{47}$$

where the superscripts (i) and (ii) refer to the particular cases (i) and (ii) [(40) and (41)] and K is the stress-intensity factor defined by

$$K = \underset{x\to 1_-}{Lt}\; [2(1-x)]^{1/2}\, \sigma_{yy}(x, 0)\,. \tag{48}$$

It is now clear that $u_y(x, 0)$ is not absolutely integrable in $-\infty < x < \infty$ since it does not vanish as $|x| \to \infty$ [cf. (43) and (46)]. It is readily verified that with the stress distribution outside the crack as obtained in (42) and (45), the boundary condition

$$\frac{\partial u_y}{\partial x}(x, 0^+) = 0 \quad \text{for} \quad |x| < 1$$

is satisfied in both the cases. Further the solution (45) is in full agreement with that derived via complex stress functions method by Erdogan [12], [13]. (To see this, one has only to superpose the solutions as given by equation (35) of [12] and (33) of [13] with $\alpha = 1$, $\gamma = 0$ for the homogeneous elastic plane and with $\omega = 0$).

6. Concluding remarks

The method of this paper is well-suited for attacking two-dimensional elastostatic boundary value problems through Sneddon's representation of the displacement vector in terms of harmonic functions and can be employed whether or not the displacement components are Fourier-transformable, provided, of course, transforms exist for their first order derivatives and hence for the components of stress. The method is more general than the usual transform method in the sense that whenever the latter is employable, this method also works. Further the method of GFT as discussed in this paper avoids the occurrence of divergent integrals in the formulation or in the solution of the usual transform method of solution. Also the unknown functions in the singular integral equations of the present formulation are those which are the most important from the practical view point, namely, the stresses outside the crack and by solving the above integral equations, these quantities are obtained directly.

To conclude, we are inclined to agree with ERDOGAN [9] that the method of integral transforms has its own limitations. However, by generalising the transforms suitably, in a manner similar to what has been done by GOLECKI [1] and what we have done in this paper for half-plane problems, the authors believe that it is possible to get over the difficulties which may be encountered in the usual transform method.

References

1 GOLECKI, J. J., Divergent integral appearing in FOURIER transform applications in elasticity, Int. J. Math. Educ. Sci. Technol. 6 pp. 321–326 (1975).
2 SNEDDON, I. N., Solutions of the equations of thermoelastic equilibrium, Archiwum Mech. Stos. 14, pp. 113–125 (1962).
3 LOVE, A. E. H., A treatise on the mathematical theory of elasticity, Dover Publications, New York, 1944.
4 SNEDDON, I. N., FOURIER transforms, McGraw-Hill Book Co., New York, 1951.
5 SNEDDON, I. N., and LOWENGRUB, M., Crack problems in the classical theory of elasticity, John Wiley & Sons, Inc., New York, 1969.
6 SIH, G. C., (Ed.), Mechanics of fracture, volume 1, Noordhoff International Publishing, Leyden, 1973.
7 MANIVACHAKAN, K.; CHAKRABARTI, A., A generalised FOURIER transform and its use in FLAMANT's problem of the classical theory of elasticity (To appear in Int. J. Math. Educ. Sci. Technol.).
8 MANIVACHAKAN, K.; CHAKRABARTI, A., Application of a generalised FOURIER transform to a crack problem ZAMM 61, pp 122–124 (1981)
9 ERDOGAN, F., Stress distribution in bonded dissimilar materials containing circular or ring-shaped cavities, J. Appl. Mech. 32 pp. 829–836 (1965).
10 ERDOGAN, F., Simultaneous dual integral equations with trigonometric and BESSEL kernels, ZAMM 48, pp. 217–225 (1968).
11 LOWENGRUB, M., Some dual trigonometric equations with an application to elasticity, Int. J. Engng. Sci. 4, pp. 69–79 (1966).
12 ERDOGAN, F., Stress distribution in a non-homogeneous elastic plane with cracks, J. Appl. Mech. 30, pp. 232–236 (1963).
13 ERDOGAN, F., Stress distribution in bonded dissimilar materials with cracks, J. Appl. Mech. 32, pp. 403–410 (1965).
14 GRADSHTEYN, I. S.; RYZHIK, I. M., Tables of integrals, series and products, Academic Press, New York, 1965.
15 MUSKHELISHVILI, N. I., Singular integral equations, P. Noordhoff, Groningen, 1953.

Eingereicht am 7. 7. 1980

Anschrift: Dr. A. CHAKRABARTI, Department of Applied Mathematics, Indian Institute of Sciences, Bangalore — 560012, India

ZAMM 61, 365–377 (1981)

M. N. Farah

Laminare, instationäre Strömung im Ringraum zwischen zwei Rohren, von denen das innere harmonisch schwingt

In dieser Arbeit wird untersucht, wie die laminare Strömung im Ringraum zwischen zwei Rohren von der harmonischen Schwingung des Innenzylinders beeinflußt wird. Im Rahmen einer Störungsrechnung zweiter Ordnung wird die Durchflußmenge und die Leistung des schwingenden Innenzylinders errechnet. Die Rechnung ergibt: a) Bei sehr langsamen Schwingungen (quasistatischer Fall) erhöht sich der Durchfluß gegenüber der konzentrischen Ruhelage. Durch wachsendes Radienverhältnis $x = r_1/r_2$ wird dieser Effekt verstärkt. Der Grenzwert $\omega \to 0$ wird mit der exakten Lösung von Cardwell verglichen. b) Für höhere Frequenzen mindert sich der Durchfluß, bis ein Minimalwert erreicht wird, der weit unter dem Durchfluß bei konzentrischer Ruhelage liegt. c) Für sehr hohe Frequenzen steigt der Durchfluß wieder allmählich an, bis er asymptotisch einen Wert erreicht, der noch unter der Ruhelage liegt. d) Die erforderliche Leistung/Länge L zur Aufrechterhaltung der vorgegebenen harmonischen Schwingung des Innenzylinders wächst mit steigender Frequenz.*

In this paper it is analysed how the laminar flow in the annulus between two tubes depends on the harmonic oscillations of the inner cylinder. In the framework of a second order perturbation theory the flow rate and the power of the oscillating inner cylinder is calculated. The results of the calculations are: a) In the case of very slow oscillations (the quasi-stationary case) the flow rate is higher than in the case where the two cylinders are coaxially and the inner cylinder is at rest. This effect is strengthened with increasing ratio $x = r_1/r_2$. The limit case $\omega \to 0$ is compared with the exact solution of Cardwell. b) With increasing frequencies the flow rate decreases down to a minimum much lower than the flow rate in the case of the coaxial steady position of the inner cylinder. c) In the case of very high frequencies the flow rate increases asymptotically up to a value lower than the value for the coaxial steady position of the inner cylinder. d) The power per length L required for preservation of the given harmonic oscillations of the inner cylinder increases with increasing frequencies.

В этой работе исследуется, как в зависимости от гармонического колебания внутренного цилиндра изменяется ламинарное течение в кольцевом пространстве между двумя трубами. В рамках нелинейной теории возмущений вычисляется расход жидкости и мощность колеблующегося внутренного цилиндра. В результате этих вычислений получается: а) При очень медленных колебаниях (квазистационарный случай) расход жидкости увеличивается по отношению к концентрическому положению покоя. С ростом отношения радиусов $x = r_1/r_2$ этот эффект усиливается. Предельный случай при $\omega \to 0$ сравнивается с точным решением Кардвелла. б) Для больших частот расход жидкости уменьшается до некоторого минимального значения, которое на много меньше значения расхода при концентрическом положении покоя. в) Для очень высоких частот расход снова медленно возрастает и асимптотически приближается к значению меньше значения в случае положения покоя. г) Потребуемая мощность по длине L для сохранения заданного гармонического колебания внутренного цилиндра увеличивается с ростом частоты.

1.1. Einleitung

Um einen ersten Einblick in das Zusammenspiel von Strömung und Schwingung zu gewinnen, wird in dieser Arbeit untersucht, wie sich der Durchfluß einer laminaren Strömung im Ringraum zwischen zwei Rohren ändert, wenn das innere Rohr senkrecht zu seiner Achse harmonisch schwingt. Die Abhängigkeit der zeitlich gemittelten Leistung von der Schwingungsfrequenz wird im Anschluß diskutiert. Das Problem der Selbstanregung ist damit natürlich noch nicht behandelt, jedoch kann die Energiebetrachtung der fremderregten Schwingung zumindest klären, ob die schwach gestörte Laminarströmung das Phänomen schon erklären kann. Sollte der Effekt auf periodischem Ablösen der Grenzschicht und/oder Bildung von periodischen Wirbelbändern beruhen, so wäre das hier zu Grunde gelegte Modell natürlich zu einfach. Nach dem Hagen-Poiseuilleschen Gesetz

$$Q = \frac{\pi R^4}{8\mu} \cdot \frac{P_1 - P_2}{l} \tag{1.1.1}$$

ist der Durchfluß (bei gleicher Druckdifferenz und gleicher Länge) durch ein Rohr genau doppelt so groß wie der Durchfluß durch zwei gleiche Rohre mit dem jeweils halben Querschnitt. Also kann man eine Erhöhung des Durchflusses allein durch Formänderung der Querschnittsfläche (bei gleicher Gesamtfläche) erwarten.

Für laminare Strömungen hat Cardwell [5] gezeigt, daß die Durchflußmenge im exzentrischen Ringspalt größer ist als im konzentrischen. Man kann deshalb für langsame Schwingungen des inneren Rohres ebenfalls eine zunehmende Durchflußmenge erwarten, während bei schnellen Schwingungen infolge der erhöhten Dissipation ein gegenteiliger Effekt zu vermuten ist.

Das Ziel dieser Arbeit besteht darin, für kleine Amplituden des schwingenden Innenrohres einen Zusammenhang zwischen Durchfluß und Schwingungsfrequenz sowie zwischen Leistung und Frequenz aufzustellen. Eine exakte Behandlung des nicht vereinfachten Problems ist aufgrund der Nichtlinearität des Konvektivbeschleunigungsgliedes in den Navier-Stokes-Gleichungen sicher sehr schwierig und wohl nur numerisch durchführbar. Deshalb wird eine Störungsrechnung durchgeführt. Als Störparameter ε bietet sich eine dimensionslose Form der Schwingungsamplitude an. Da aus Symmetriegründen die instationäre Durchflußänderung $Q - Q_0 = \varepsilon^2 Q_2$ erst von der quadratischen Störpotenz abhängen kann, ist eine Theorie zweiter Ordnung erforderlich.

*) Großes Radienverhältnis: enger Spalt; Kleines Radienverhältnis: großer Spalt

2.1 Grundgleichungen

Mit den Navier-Stokes-Gleichungen

$$\frac{D\boldsymbol{u}}{Dt} = \frac{\partial \boldsymbol{u}}{\partial t} + (\boldsymbol{u} \cdot \nabla)\,\boldsymbol{u} = -\frac{1}{\varrho}\nabla P + \nu\Delta\boldsymbol{u} \tag{2.1.1}$$

und der Kontinuitätsgleichung

$$\nabla \cdot \boldsymbol{u} = 0 \tag{2.1.2}$$

erhält man vier Bestimmungsgleichungen für die vier unbekannten Größen P und $\boldsymbol{u} = (u, v, w)$. Die Dichte ϱ und die kinematische Zähigkeit $\nu = \mu/\varrho$ werden als konstant vorausgesetzt. Für das Folgende ist es zweckmäßiger, die Komponente in Achsrichtung gesondert zu behandeln (Bild 1)

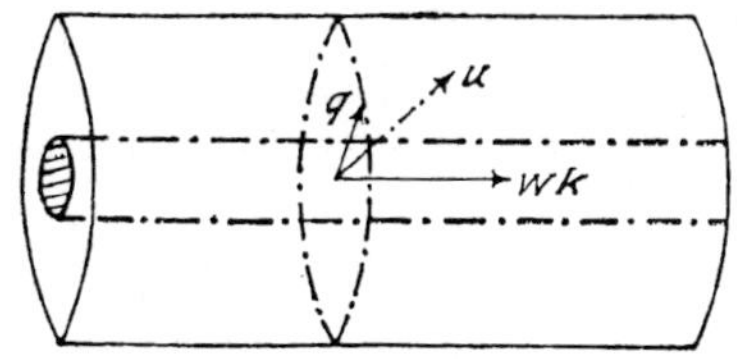

Bild 1

$$\boldsymbol{u} = \boldsymbol{q} + w\boldsymbol{k}\,, \qquad \boldsymbol{q} = u\boldsymbol{i} + v\boldsymbol{j}\,. \tag{2.1.3}$$

Wird die Strömung als voll ausgebildet angenommen,

$$\boldsymbol{u} = \boldsymbol{u}(x, y, t) \Rightarrow P = -C \cdot z + \tilde{P}(x, y, t)\,, \tag{2.1.4}$$

so folgt aus (2.1.1—4) mit der Vereinbarung

$$\operatorname{grad} \tilde{P} = \frac{\partial \tilde{P}}{\partial x}\boldsymbol{i} + \frac{\partial \tilde{P}}{\partial y}\boldsymbol{j}; \qquad \operatorname{div} \boldsymbol{q} = \frac{\partial u}{\partial x} + \frac{\partial v}{\partial y} \tag{2.1.5}$$

das System von Differentialgleichungen

$$\frac{\partial \boldsymbol{q}}{\partial t} + (\boldsymbol{q} \cdot \operatorname{grad})\,\boldsymbol{q} = -\frac{1}{\varrho}\operatorname{grad} \tilde{P} + \nu\,\Delta\boldsymbol{q}\,, \tag{2.1.6}$$

$$\operatorname{div} \boldsymbol{q} = 0\,,$$

$$\frac{\partial w}{\partial t} + (\boldsymbol{q} \cdot \operatorname{grad})\,w = -\frac{C}{\varrho} + \nu\,\Delta w\,.$$

Durch Einführung einer Stromfunktion für die Querströmung

$$\boldsymbol{q} = \boldsymbol{k} \times \operatorname{grad} \psi \tag{2.1.7}$$

wird die Kontinuitätsgleichung identisch befriedigt. Für die Stromfunktion schreibt sich die Gleichung der Querströmung wie folgt:

$$\left(\frac{\partial}{\partial t} - \nu\Delta\right)\Delta\psi = \left(\psi_y \frac{\partial}{\partial x} - \psi_x \frac{\partial}{\partial y}\right)\Delta\psi\,. \tag{2.1.8}$$

2.2 Randbedingungen

Zunächst wird der Störparameter

$$\varepsilon = \lambda/r_1\,, \qquad r_1\text{: Innenradius}, \lambda\text{: Schwingungsamplitude} \tag{2.2.1}$$

definiert. Der geometrische Ort der Oberfläche des schwingenden Innenrohres lautet somit

$$\boldsymbol{r}_\varepsilon = \boldsymbol{r}_1 + \varepsilon r_1 \sin \omega t \boldsymbol{i}\,. \tag{2.2.2}$$

Die Haftbedingungen am inneren Rohr ergeben sich daraus zu

$$w(\varepsilon, t) = 0\,, \qquad \boldsymbol{q}(\boldsymbol{r}_\varepsilon, t) = \varepsilon r_1 \omega \cos \omega t \boldsymbol{i}\,. \tag{2.2.3}$$

Das äußere Rohr bleibt während des Vorganges in Ruhe:

$$w(\boldsymbol{r}_2, t) = 0\,, \qquad \boldsymbol{q}(\boldsymbol{r}_2, t) = 0\,. \tag{2.2.4}$$

Im Folgenden setzen wir die Existenz einer laminaren instationären Strömung des vollen nichtlinearen Randwertproblems (2.1.6—8)/(2.2.3—4) voraus.

3.1 Theorie kleiner Störungen

Die als existent vorausgesetzte Lösung $\psi(x, y, t; \varepsilon)$ hängt außer von den Ortskoordinaten auch vom Parameter ε ab. Wir setzen voraus, daß diese Abhängigkeit analytisch ist, und können dann nach Potenzen des Störparameters entwickeln:

$$\psi = \varepsilon\psi_1 + \varepsilon^2\psi_2 + \dots\,, \qquad w = w_0 + \varepsilon w_1 + \varepsilon^2 w_2 + \dots \tag{3.1.1}$$

Die Koeffizienten der Störpotenzen ε^n müssen (nach erfolgter Entwicklung und Umordnung) in den Erhaltungssätzen identisch verschwinden. Das bedingt das Differentialgleichungssystem nullter, erster, zweiter, ... Ordnung.

3.2 Die projizierten Randbedingungen

Für die folgende Überlegung denke man sich die nichtlineare Lösung über den Innenrand fortgesetzt. Bei genügend glattem Rand ist die Existenz einer Fortsetzung bis hin zu der konzentrischen Ausgangslage des Innenzylinders für genügend kleine Amplituden keine sehr einschneidende Forderung. Die Störkomponenten in den Randbedingungen

$$\varepsilon r_1\omega \cos \omega t\boldsymbol{i} = \varepsilon \boldsymbol{q}_1(\boldsymbol{r}_\varepsilon) + \varepsilon^2 \boldsymbol{q}_2(\boldsymbol{r}_\varepsilon) + \dots, \qquad 0 = w_0(\boldsymbol{r}_\varepsilon) + \varepsilon w_1(\boldsymbol{r}_\varepsilon) + \varepsilon^2 w_2(\boldsymbol{r}_\varepsilon) + \dots \tag{3.2.1}$$

(mit $\boldsymbol{r}_\varepsilon = \boldsymbol{r}_1 + \varepsilon r_1 \sin \omega t\boldsymbol{i}$) werden nun ebenfalls in eine TAYLORreihe bezüglich ε um $\varepsilon = 0$ entwickelt. Koeffizientenvergleich führt daraufhin zu den gesuchten projizierten Randbedingungen am starren Innenzylinder.

3.3 Die stationäre Grundströmung

Aus Symmetriegründen hängt die Grundströmung w_0 nur vom Radius r ab. Das allgemeine Integral der Differentialgleichung

$$\frac{1}{r}\frac{\mathrm{d}}{\mathrm{d}r}\left(r\frac{\mathrm{d}w_0}{\mathrm{d}r}\right) = -\frac{C}{u} \Rightarrow w_0 = -\frac{C}{4\mu}r^2 + C_1 \ln r + C_2 \tag{3.3.1}$$

wird an die Randbedingungen

$$w_0(r_1) = 0; \qquad w_0(r_2) = 0 \tag{3.3.2}$$

angepaßt und ergibt das stationäre Strömungsfeld im Ringspalt

$$w_0 = \frac{C}{4\mu}\left[(r_2^2 - r_1^2)\frac{\ln r/r_1}{\ln r_2/r_1} - (r^2 - r_1^2)\right]. \tag{3.3.3}$$

4.1 Bemerkungen zur allgemeinen Lösungsmethode

In den folgenden Randwertproblemen treten überhaupt nur zwei Typen von Differentialgleichungen auf:

$$\left(\frac{\partial}{\partial t} - \nu\Delta\right) w = \beta; \qquad \left(\frac{\partial}{\partial t} - \nu\Delta\right)\Delta\psi = \beta^* . \tag{4.1.1}$$

Aufgrund der speziellen Struktur der hier auftretenden Randwertprobleme können diese partiellen Differentialgleichungen auf gewöhnliche reduziert werden. Wir untersuchen zunächst die homogenen Gleichungen ($\beta = \beta^* = 0$). Aus der Vertauschbarkeit der linearen Operatoren $\left(\frac{\partial}{\partial t} - \nu\Delta\right)$ und Δ folgt, (siehe HOLTSMARK [6]), daß sich die homogene Lösung der zweiten Gleichung (4.1.1) additiv zusammensetzen läßt:

$$\psi_h = \psi_1 + \psi_2 \quad \text{mit} \quad \Delta\psi_1 = 0 \quad \text{und} \quad \left(\frac{\partial}{\partial t} - \nu\Delta\right)\psi_2 = 0 . \tag{4.1.2}$$

Es genügt also vollauf, die beiden Gleichungen

$$\Delta f = 0; \qquad \left(\frac{\partial}{\partial t} - \nu\Delta\right) g = 0 \tag{4.1.3}$$

einzeln zu behandeln. Da eine harmonische Schwingung nur ganzzahlige Vielfache der Grundfrequenz anregen kann und da sowohl w als auch ψ bei einem Umlauf um den inneren Zylinder stetig sein müssen, genügt (wegen des Superpositionsprinzips linearer Gleichungen) die Untersuchung der folgenden Separationsansätze in Polarkoordinaten

$$f_{n,m} = F_n(r)\,\mathrm{e}^{jn\Theta}\,\mathrm{e}^{im\omega t}, \qquad g_{n,m} = G_{n,m}(r)\,\mathrm{e}^{jn\Theta}\,\mathrm{e}^{im\omega t}; \qquad n, m \in N \tag{4.1.4}$$

(Bemerkung: Die unterschiedliche Schreibweise von $j = \sqrt{-1}$ und $i = \sqrt{-1}$ ist nur ein technischer Trick, um das vollständige Fundamentalsystem einfacher darstellen zu können). Mit diesem Ansatz reduzieren sich die LAPLACE- und die Wärmeleitungsgleichung (4.1.3) auf EULERsche und (modifizierte) BESSELsche Differentialgleichungen:

$$\left[\frac{\mathrm{d}^2}{\mathrm{d}r^2} + \frac{1}{r}\frac{\mathrm{d}}{\mathrm{d}r} - \frac{n^2}{r^2}\right] F_n(r) = 0, \qquad \left[\frac{\mathrm{d}^2}{\mathrm{d}r^2} + \frac{1}{r}\frac{\mathrm{d}}{\mathrm{d}r} - \left(\frac{n^2}{r^2} + \frac{im\omega}{\nu}\right)\right] G_{n,m}(r) = 0 . \tag{4.1.5}$$

Um ständige Wiederholungen zu vermeiden, führen wir ein Symbol für den (komplexen) BESSELschen Differentialoperator ein:

$$\mathcal{B}_{n,m} := \frac{\mathrm{d}^2}{\mathrm{d}r^2} + \frac{1}{r}\frac{\mathrm{d}}{\mathrm{d}r} - \left(\frac{n^2}{r^2} + \frac{im\omega}{\nu}\right). \tag{4.1.6}$$

Das Fundamentalsystem der Differentialgleichungen (4.1.5)

$$F_n = C_1 r^n + C_2 r^{-n} \quad (n \neq 0), \qquad F_0 = C_1 \ln r + C_2 , \tag{4.1.7}$$

$$G_{n,m} = C_1 I_n\left(\sqrt{\frac{im\omega}{\nu}}\, r\right) + C_2 K_n\left(\sqrt{\frac{im\omega}{\nu}}\, r\right)$$

ist nur für die F_n reell. Für die rechnerische Auswertung müssen die Zylinderfunktionen (die wir generell mit Z_n abkürzen wollen),

$$I_n(z) = \frac{1}{\pi} \int_0^\pi \exp(z \cos\tau) \cos(n\tau)\, d\tau; \qquad K_n(z) = \int_0^\infty \exp(-z \cosh\tau) \cosh(n\tau)\, d\tau, \tag{4.1.8}$$

wieder in Real- und Imaginärteil zerlegt werden. In der Literatur (siehe McLACHLAN [10]) werden hierfür die normierten KELVIN-Funktionen eingeführt (x: reell):

$$e^{(n/2)\pi i} I_n(e^{i(\pi/4)} x) = \mathrm{ber}_n(x) + i\, \mathrm{bei}_n(x), \qquad e^{-(n/2)\pi i} K_n(e^{i(\pi/4)} x) = \mathrm{ker}_n(x) + i\, \mathrm{kei}_n(x), \tag{4.1.9}$$

die vollständig vertafelt vorliegen. Das Fundamentalsystem in komplexer Schreibweise [11]

$$f_{n,m} = r^{\pm n}\, e^{(jn\Theta + im\omega t)}, \qquad g_{n,m} = Z_n\left(\sqrt{\frac{im\omega}{\nu}}\, r\right) e^{(jn\Theta + im\omega t)} \tag{4.1.10}$$

verkörpert also insgesamt jeweils acht verschiedene Elementarlösungen. An dem folgenden Beispiel wird das Auffinden der reellen Lösungen exemplarisch demonstriert:

$$\mathrm{Re}_i\, \mathrm{Im}_j \{(\mathrm{ker}(x) + i\, \mathrm{kei}(x))\, (\cos(\Omega) + i \sin(\Omega))\, (\cos(\Theta) + j \sin(\Theta))\} =$$
$$= (\mathrm{ker}(x) \cos(\Omega) - \mathrm{kei}(x) \sin(\Omega)) \sin(\Theta). \tag{4.1.11}$$

Mit der Kenntnis des Fundamentalsystems kann eine partikuläre Lösung einer inhomogenen Gleichung sofort angegeben werden (Variation der Konstanten) [7].

4.2 Die Querströmung in linearer Näherung

Man formuliert dieses Randwertproblem in der folgenden Form

$$\left(\frac{\partial}{\partial t} - \nu\Delta\right) \Delta\psi_1 = 0; \qquad \boldsymbol{q}_1 = \boldsymbol{k} \times \Delta\psi_1; \qquad \boldsymbol{q}_1(r_1) = r_1\omega \cos\omega t\boldsymbol{i}; \qquad \boldsymbol{q}_1(r_2) = 0. \tag{4.2.1}$$

Aus der Randbedingung folgt, daß ψ_1 die folgende Gestalt besitzen muß:

$$\psi_1 = \underset{i}{\mathrm{Re}}\, \underset{j}{\mathrm{Im}} \{F(r)\, e^{j\Theta + i\omega t}\}. \tag{4.2.2}$$

Die komplexe Funktion F genügt also der Differentialgleichung $\underset{1,1}{\mathscr{B}}\underset{1,0}{\mathscr{B}}F(r) = 0$ mit der Lösung:

$$F = C_1 r + C_2 \frac{1}{r} + C_3 I_1\left(\sqrt{\frac{i\omega}{\nu}}\, r\right) + C_4 K_1\left(\sqrt{\frac{i\omega}{\nu}}\, r\right). \tag{4.2.3}$$

Für spätere Zwecke wird die Zerlegung in den harmonischen und nichtharmonischen Anteil bereitgestellt:

$$\begin{aligned} \psi_1 &= \psi_{11} + \psi_{12}, \\ \psi_{11} &= \underset{i}{\mathrm{Re}}\, \underset{j}{\mathrm{Im}} \left\{C_1 r + C_2 \frac{1}{r}\right\} e^{j\Theta + i\omega t}, \\ \psi_{12} &= \underset{i}{\mathrm{Re}}\, \underset{j}{\mathrm{Im}} \left\{C_3 I_1\left(\sqrt{\frac{i\omega}{\nu}}\, r\right) + C_4 K_1\left(\sqrt{\frac{i\omega}{\nu}}\, r\right)\right\} e^{j\Theta + i\omega t}. \end{aligned} \tag{4.2.4}$$

Die vier Randbedingungen

$$\begin{aligned} &\frac{\partial\psi_1}{\partial r}(r_1) = -r_1\omega \sin\Theta \cos\omega t; \qquad \frac{\partial\psi_1}{\partial r}(r_2) = 0 \\ &\frac{\partial\psi_1}{\partial\Theta}(r_1) = -r_1^2\omega \cos\Theta \cos\omega t; \qquad \frac{\partial\psi_1}{\partial\Theta}(r_2) = 0 \end{aligned} \tag{4.2.5}$$

legen die komplexen Konstanten C_n fest. Es wird behauptet, daß sich durch das lineare Gleichungssystem

$$F'(r_1) = -r_1\omega, \qquad F(r_1) = -r_1^2\omega, \qquad F'(r_2) = 0, \qquad F(r_2) = 0 \tag{4.2.6}$$

die korrekten Konstanten ergeben. Der Beweis wird exemplarisch an der ersten Randbedingung (4.2.5) vorgeführt:

$$\frac{\partial}{\partial r} \underset{i}{\mathrm{Re}}\, \underset{j}{\mathrm{Im}} \{F(r)\, e^{j\Theta + i\omega t}\}|_{r=r_1} = \underset{i}{\mathrm{Re}}\, \underset{j}{\mathrm{Im}} \{F'(r)\, e^{j\Theta + i\omega t}\}|_{r=r_1} = (-r_1\omega)\, \underset{i}{\mathrm{Re}}\, \underset{j}{\mathrm{Im}} \{e^{i\Theta + i\omega t}\}. \tag{4.2.7}$$

Unter Ausnutzung der WRONSKIdeterminante und den Ableitungsrelationen

$$I_n(z)\, K_n'(z) - K_n(z)\, I_n'(z) = \frac{1}{z}, \qquad I_n'(z) = I_{n+1}(z) + \frac{n}{z} I_n(z), \qquad K'_n(z) = -K_{n+1}(z) + \frac{n}{z} K_n(z) \tag{4.2.8}$$

berechnen sich die vier Konstanten zu

$$C_1 = (I_{02}K_{21} - I_{21}K_{02} - 2/i\varepsilon_1^2)\cdot(\omega r_1)/(r_2^2 D)\,,\qquad C_2 = (I_{21}K_{22} - I_{22}K_{21})\cdot(\omega r_1)/D\,,$$
$$C_3 = (K_{22}/r_1^2 - K_{21}/r_2^2)\cdot(2\omega r_1^2)/(\sqrt{i}\,\varepsilon_1 D)\,,\qquad C_4 = (I_{22}/r_1^2 - I_{21}/r_2^2)\cdot(2\omega r_1^2)/(\sqrt{i}\,\varepsilon_1 D)\,, \tag{4.2.9}$$
$$D = (I_{22}K_{01} - I_{01}K_{22})/r_1^2 - (I_{02}K_{21} - I_{21}K_{02})/r_2^2 + 4/(i\varepsilon_1^2 r_2^2)\,.$$

Die Kurzschreibweise bedeutet dabei

$$Z_{n1} = Z_n(\sqrt{i}\,\varepsilon_1);\qquad Z_{n2} = Z_n\left(\sqrt{i}\,\varepsilon_1\frac{r_2}{r_1}\right);\qquad \varepsilon_1 = \sqrt{\frac{\omega}{\nu}}\,r_1\,. \tag{4.2.10}$$

Für spätere Zwecke wird noch der Querströmungsvektor im zylindrischen Basissystem angegeben:

$$\boldsymbol{q}_1 = \frac{\partial\psi_1}{\partial r}\boldsymbol{e}_\Theta - \frac{1}{r}\frac{\partial\psi_1}{\partial\Theta}\boldsymbol{e}_r = \underset{i}{\mathrm{Re}}\,\underset{j}{\mathrm{Im}}\left\{\left(F'\,\boldsymbol{e}_\Theta - j\frac{F}{r}\boldsymbol{e}_r\right)e^{j\Theta+i\omega t}\right\}. \tag{4.2.11}$$

4.3 Die Längsströmung in linearer Näherung

Der Quellterm des linearen Längsströmungsproblems ist mit (4.2.11)/(3.3.3) als bekannt anzusehen:

$$\left(\frac{\partial t}{\partial} - \nu\Delta\right)w_1 = -(\boldsymbol{q}_1\cdot\nabla)\,w_0;\qquad w_1(r_1) = -r_1\sin\omega t\frac{\partial w_0}{\partial x};\qquad w_1(r_2) = 0\,. \tag{4.3.1}$$

Da w_0 nur von r abhängt, vereinfacht sich die rechte Seite zu

$$-(\boldsymbol{q}_1\cdot\nabla)\,w_0 = \underset{i}{\mathrm{Re}}\,\underset{j}{\mathrm{Re}}\left\{\frac{F(r)}{ir}\cdot w_0'(r)\,e^{j\theta+i\omega t}\right\}. \tag{4.3.2}$$

Weil die Randbedingung am Innenzylinder

$$w_1(r_1) = \underset{i}{\mathrm{Re}}\,\underset{j}{\mathrm{Re}}\,\{ir_1 w_0'(r_1)\,e^{j\Theta+i\omega t}\} \tag{4.3.3}$$

sich ebenfalls in der gleichen Weise darstellen läßt, gelingt die Reduktion auf ein einfaches Randwertproblem

$$w_1 = \underset{i}{\mathrm{Re}}\,\underset{j}{\mathrm{Re}}\,\{G(r)\,e^{j\Theta+i\omega t}\}\,,$$
$$\mathcal{B}_{1,1}G(r) = -\frac{1}{\nu r}F(r)\,w_0'(r) = Q(r)\,, \tag{4.3.4}$$
$$G(r_1) = ir_1 w_0'(r_1)\,,\qquad G(r_2) = 0\,.$$

Die aus dem homogenen und partikulären Anteil zusammengesetzte Lösung

$$G = G_h + G_p\,,$$
$$G_h = C_5 I_1(\tau r) + C_6 K_1(\tau r);\qquad \tau = \sqrt{\frac{i\omega}{\nu}}\,, \tag{4.3.5}$$
$$G_p = -\int^r [I_1(\tau\varrho)\,K_1(\tau r) - K_1(\tau\varrho)\,I_1(\tau r)]\,Q(\varrho)\,\varrho\,d\varrho$$

muß noch an die Randbedingungen angepaßt werden. Aus dem linearen Gleichungssystem

$$G_h(r_1) = ir_1 w_0'(r_1) - G_p(r_1) = Q_1(r_1)\,,\qquad G_h(r_2) = -G_p(r_2) = Q_2(r_2) \tag{4.3.6}$$

folgen die gesuchten Konstanten

$$C_5 = \begin{vmatrix} Q_1 & K_{11}\\ Q_2 & K_{12}\end{vmatrix}\Big/D;\qquad C_6 = \begin{vmatrix} I_{11} & Q_1\\ I_{12} & Q_2\end{vmatrix}\Big/D;\qquad D = \begin{vmatrix} I_{11} & K_{11}\\ I_{12} & K_{12}\end{vmatrix}. \tag{4.3.7}$$

Der schwierigste Teil des skizzierten Rechenwegs ist die Integration des partikulären Integrals G_p in (4.3.5). Die Integrale führen leider nur zum Teil wieder auf Ausdrücke der Form $r^m Z_n$ oder $r^m Z_n Z_K$. Da auch neue Klassen von speziellen Funktionen (LOMMEL-Funktionen; siehe LUKE [8]) auftreten, ist es bequemer, die Integraldarstellung (4.3.5) beizubehalten.

4.4 Die Querströmung zweiter Ordnung

Das Querströmungsproblem zweiter Ordnung lautet

$$\left(\frac{\partial}{\partial t} - \nu\Delta\right)\Delta\psi_2 = -(\boldsymbol{q}_1\cdot\nabla)\,\Delta\psi_1;\qquad \boldsymbol{q}_2 = \boldsymbol{K}\times\nabla\psi_2\,, \tag{4.4.1}$$
$$\boldsymbol{q}_2(r_1) = -r_1\sin\omega t\frac{\partial\boldsymbol{q}_1}{\partial x};\qquad \boldsymbol{q}_2(r_2) = 0\,.$$

Da ein Teil der homogenen Lösung von ψ_1 nach (4.2) selbst schon der LAPLACEgleichung genügt,

$$\psi_1 = \psi_{11} + \psi_{12}; \qquad \Delta\psi_{11} = 0; \qquad \left(\frac{\partial}{\partial t} - \nu\Delta\right)\psi_{12} = 0\,, \tag{4.4.2}$$

vereinfacht sich der Quellterm zu

$$-(\boldsymbol{q}_1 \cdot \nabla)\,\Delta\psi_{11} = -\frac{1}{\nu r}\left(\frac{\partial}{\partial r}\psi_1\frac{\partial}{\partial\Theta} - \frac{\partial}{\partial\Theta}\psi_1\frac{\partial}{\partial r}\right)\frac{\partial}{\partial t}\psi_{12} = \underset{i}{\mathrm{Re}}\,\underset{j}{\mathrm{Im}}\,\{M(r)\,\mathrm{e}^{2(j\Theta+i\omega t)} + N(r)\,\mathrm{e}^{2j\Theta}\}\,. \tag{4.4.3}$$

Hierin bedeuten M und N die neu eingeführten komplexen Funktionen

$$M(r) = -\frac{i\omega}{4\nu}\,\frac{1}{r}\,(F_1F_2' - F_2F_1'); \qquad N(r) = -\frac{i\omega}{4\nu}\,\frac{1}{r}\,(F\bar{F}_2' + F_2\bar{F}') \tag{4.4.4}$$

mit der zu (4.4.2) entsprechenden Zerlegung $F(r) = F_1(r) + F_2(r)$; $\mathcal{B}_{1,0}F_1 = 0$; $\mathcal{B}_{11}F_2 = 0$. Der Ausdruck $\bar{F}$ symbolisiere den konjugierten Funktionswert zu F.

Das Problem reduziert sich diesmal auf zwei inhomogene Differentialgleichungen

$$\psi_2 = \underset{i}{\mathrm{Re}}\,\underset{j}{\mathrm{Im}}\,\{H(r)\,\mathrm{e}^{2(j\Theta+i\omega t)} + L(r)\,\mathrm{e}^{2j\Theta}\}\,, \tag{4.4.5}$$

$$\mathcal{B}_{2,2}\mathcal{B}_{2,0}\,H(r) = -\frac{1}{\nu}M(r)\,, \qquad \mathcal{B}_{2,0}\mathcal{B}_{2,0}\,L(r) = -\frac{1}{\nu}N(r)$$

mit den jeweiligen Randbedingungen

$$\begin{aligned}
H(r_1) &= \frac{ir_1}{4}\left(\frac{\mathrm{d}}{\mathrm{d}r} - \frac{1}{r}\right)F|_{r_1}; \qquad H(r_2) = 0\,,\\
H'(r_1) &= \frac{ir_1}{4}\left(\frac{\mathrm{d}^2}{\mathrm{d}r^2} - \frac{1}{r}\,\frac{\mathrm{d}}{\mathrm{d}r} + \frac{1}{r^2}\right)F|_{r_1}; \qquad H'(r_2) = 0\,,\\
L(r_1) &= \frac{ir_1}{4}\left(\frac{\mathrm{d}}{\mathrm{d}r} - \frac{1}{r}\right)\bar{F}|_{r_1}; \qquad L(r_2) = 0\,,\\
L'(r_1) &= \frac{ir_1}{4}\left(\frac{\mathrm{d}^2}{\mathrm{d}r^2} - \frac{1}{r}\,\frac{\mathrm{d}}{\mathrm{d}r} + \frac{1}{r^2}\right)\bar{F}|_{r_1}; \qquad L'(r_2) = 0\,.
\end{aligned} \tag{4.4.6}$$

Die Lösungsmethode ist in (4.1) skizziert. Da es sich herausstellt, daß für den zeitlich gemittelten Massendurchfluß die Kenntnis der Querströmung zweiter Ordnung nicht erforderlich ist, soll die bloße Formulierung hier genügen.

4.5 Die Längsströmung zweiter Ordnung

Zum Schluß wird noch die Längsströmung zweiter Ordnung soweit behandelt, wie es die Themenstellung erfordert.

$$\begin{aligned}
&\left(\frac{\partial}{\partial t} - \nu\Delta\right)w_2 = -(\boldsymbol{q}_1 \cdot \nabla)\,w_1 - (\boldsymbol{q}_2 \cdot \nabla)\,w_0\,,\\
&w_2(r_1) = -r_1 \sin\omega t\,\frac{\partial w_1}{\partial x} - \frac{1}{2}\,r_1 \sin^2\omega t\,\frac{\partial^2 w_0}{\partial x^2}\,; \qquad w_2(r_2) = 0\,.
\end{aligned} \tag{4.5.1}$$

Die komplexe Formulierung führt jetzt auf vier Differentialgleichungen

$$\begin{aligned}
&W_2 = \underset{i}{\mathrm{Re}}\,\underset{j}{\mathrm{Re}}\,\{R(r)\,\mathrm{e}^{2(j\Theta+i\omega t)} + S(r)\,\mathrm{e}^{2j\Theta} + T(r)\,\mathrm{e}^{2i\omega t} + U(r)\}\,,\\
&\mathcal{B}_{2,2}\,R(r) = -\frac{1}{\nu}\,V(r)\,, \qquad \mathcal{B}_{2,0}\,S(r) = -\frac{1}{\nu}\,W(r)\,, \qquad \mathcal{B}_{0,2}\,T(r) = -\frac{1}{\nu}\,X(r)\,,\\
&\mathcal{B}_{0,0}\,U(r) = -\frac{1}{\nu}\,Y(r)\,,\\
&V(r) = \frac{1}{4r}(G'F - GF') + \frac{2}{r}Hw_0'\,, \qquad W(r) = \frac{1}{4r}(G'\bar{F} - G\bar{F}') + \frac{2}{r}LW_0'\,,\\
&X(r) = \frac{1}{4r}(G'F + GF')\,, \qquad Y(r) = \frac{1}{4r}(G'\bar{F} + G\bar{F}')
\end{aligned} \tag{4.5.2}$$

mit den Randbedingungen

$$\begin{aligned}
&R(r_1) = Z(r_1); \qquad R(r_2) = 0\,, \qquad S(r_1) = -Z(r_1); \qquad S(r_2) = 0\,,\\
&T(r_1) = Z^*(r_1); \qquad T(r_2) = 0\,, \qquad U(r_1) = -Z^*(r_1); \qquad U(r_2) = 0\,,\\
&Z(r) = \frac{r_1^2}{8}\left(2i\left(\frac{\mathrm{d}}{\mathrm{d}r} - \frac{1}{r}\right)G + \left(\frac{\mathrm{d}^2}{\mathrm{d}r^2} - \frac{1}{r}\,\frac{\mathrm{d}}{\mathrm{d}r}\right)W_0\right),\\
&Z^*(r) = \frac{r_1^2}{8}\left(2i\left(\frac{\mathrm{d}}{\mathrm{d}r} + \frac{1}{r}\right)G + \left(\frac{\mathrm{d}^2}{\mathrm{d}r^2} + \frac{1}{r}\,\frac{\mathrm{d}}{\mathrm{d}r}\right)W_0\right).
\end{aligned} \tag{4.5.3}$$

Im nächsten Schnitt wird sich zeigen, daß der Durchfluß im zeitlichen Mittel ausschließlich von $U(r)$ abhängt. Deshalb wird nur dieses spezielle Randwertproblem weiter verfolgt. Die Differentialgleichung ist glücklicherweise von der leichtintegrierbaren Gestalt

$$\frac{1}{r}\frac{\mathrm{d}}{\mathrm{d}r}\left(r\frac{\mathrm{d}}{\mathrm{d}r}\right)U = -\frac{1}{4\nu}\frac{1}{r}\frac{\mathrm{d}}{\mathrm{d}r}(\bar{F}G) \tag{4.5.4}$$

mit der Lösung

$$U = U_h + U_p = C_7 + C_8 \ln r - \frac{1}{4\nu}\int^r \bar{F}G\frac{\mathrm{d}\varrho}{\varrho} \tag{4.5.5}$$

und den Konstanten

$$C_7 = \frac{U_p(r_2)\ln r_1 - \left(Z^*(r_1) + U_p(r_1)\right)\ln r_2}{\ln r_2/r_1}, \qquad C_8 = \frac{Z^*(r_1) + U_p(r_1) - U_p(r_2)}{\ln r_2/r_1}. \tag{4.5.6}$$

Die untere Integrationsgrenze des partikulären Integrals war willkürlich. Wir vereinbaren jetzt, daß in allen Partikularlösungen von r_1 ab integriert wird. Wegen $U_p(r_1) = 0$ vereinfacht sich (4.5.6) noch geringfügig.

5.1 Der zeitlich gemittelte Durchfluß

Es sei Q die über eine Periode T gemittelte gesamte Durchflußmenge durch die ringförmige Fläche A zwischen den beiden Zylindern:

$$Q = \frac{1}{T}\int_0^T \mathrm{d}t \iint_A w\,\mathrm{d}A; \qquad T = \frac{2\pi}{\omega}. \tag{5.1.1}$$

Da für w Analytizität bezüglich der dimensionslosen Amplitude ε vorausgesetzt wurde, ist (bei glatter Berandung) auch $Q(\varepsilon)$ analytisch und kann deshalb entwickelt werden:

$$Q = Q_0 + \varepsilon Q_1 + \varepsilon^2 Q_2 + \dots \tag{5.1.2}$$

Die Abhängigkeit $Q(\varepsilon)$ resultiert jedoch nicht nur aus der Abhängigkeit $w(\varepsilon)$, sondern auch aus der zeitlichen Änderung der Integrationsfläche $A(\varepsilon, t)$. In drei Schritten werden wir deshalb die Durchflüsse höherer Ordnung entwickeln:

1. Zu beliebiger Längsgeschwindigkeit w und beliebiger (glatter) Kontur der Innen- und Außenzylinder wird die Abhängigkeit des Durchflusses von der Schwingungsamplitude ε angegeben.
2. In dieser Darstellungsformel wird die Reihenentwicklung für $w(\varepsilon)$ eingesetzt, und der entstehende Ausdruck wird daraufhin nach Potenzen von ε geordnet.
3. Durch Berücksichtigung der Rechts-Links-Symmetrie wird die Darstellung vereinfacht.

Zu Punkt 1.): Der innere (nicht notwendig kreisförmige) Zylinder möge in x-Richtung periodisch (nicht notwendig harmonisch) schwingen. Ein Punkt der Berandung bewege sich dabei nach dem Gesetz $\tilde{x}(t) = x_0 + \varepsilon\Omega(t)$. Das Flächenintegral über A wird zerlegt in

$$\iint_A W\,\mathrm{d}A = \iint_{A_0} W\,\mathrm{d}A + \iint_{\Delta A_1} W\,\mathrm{d}A - \iint_{\Delta A_2} W\,\mathrm{d}A \tag{5.1.3}$$

ein Integral über die Fläche A_0 des ruhenden Ausgangszustandes und über die beiden „Halbmöndchen" ΔA_1 und ΔA_2; siehe Bild 2.

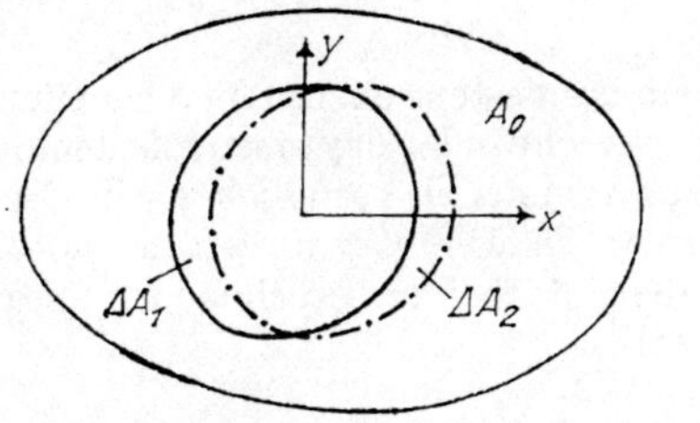

Bild 2

Die Integrale über ΔA_1 und ΔA_2,

$$f(\varepsilon) = \int_{y_u}^{y_o} \mathrm{d}y \int_{x_1(y)}^{x_1+\varepsilon\Omega} \mathrm{d}x\, W - \int_{y_u}^{y_o} \mathrm{d}y \int_{x_2(y)}^{x_2+\varepsilon\Omega} \mathrm{d}x\, W \tag{5.1.4}$$

werden nach ε entwickelt. Dies führt (mit ∂A_i: Innenzylinder) zu der gesuchten Darstellungsformel

$$\iint W\,\mathrm{d}A = \iint_{A_0} W\,\mathrm{d}A - \varepsilon\Omega \oint_{\partial A_i} W\,\mathrm{d}y - \frac{1}{2}\varepsilon^2\Omega^2 \oint_{\partial A_i} \frac{\partial W}{\partial x}\mathrm{d}y - \dots \tag{5.1.5}$$

Zu Punkt 2.: Die Integranden in (5.1.5) werden in der Form $W = W_0 + \varepsilon W_1 + \varepsilon^2 W_2 + \dots$ entwickelt, und der gesamte Ausdruck wird nach Potenzen von ε geordnet:

$$\begin{aligned}\iint_A W\,\mathrm{d}A &= \iint_{A_0} W_0\,\mathrm{d}A + \varepsilon \iint_{A_0} W_1\,\mathrm{d}A + \varepsilon^2 \iint_{A_0} W_2\,\mathrm{d}A + \dots \\ &\quad -\varepsilon\Omega \oint_{\partial A_i} W_0\,\mathrm{d}y - \varepsilon^2\Omega \oint_{\partial A_i} W_1\,\mathrm{d}y - \dots \\ &\quad -\frac{1}{2}\varepsilon^2\Omega^2 \oint_{\partial A_i} \frac{\partial W_0}{\partial x}\,\mathrm{d}y - \dots \end{aligned} \tag{5.1.6}$$

Berücksichtigt man die Randbedingungen am Innenzylinder $W_0(\partial A_i) = 0$; $W_1(\partial A_i) + \Omega \dfrac{\partial W_0}{\partial X}(\partial A_i) = 0$; so lassen sich die Konturintegrale (bis zur zweiten Ordnung) auf ein Integral zurückführen:

$$\Sigma \oint = \frac{\varepsilon^2}{2}\Omega^2 \oint_{\partial A_i} \frac{\partial W_0}{\partial x}\,\mathrm{d}y + O(\varepsilon^2)\,. \tag{5.1.7}$$

Zu Punkt 3.: Wir setzen jetzt (bezüglich x) Rechts-Links-Symmetrie voraus. Der Durchfluß (auch der zeitabhängige) muß aus Symmetriegründen eine gerade Funktion von ε sein. Also verschwinden alle ungeraden Potenzen, und wir erhalten die Ausgangsformel

$$\iint_A W\,\mathrm{d}A = \iint_{A_0} W_0\,\mathrm{d}A + \varepsilon^2\left[\iint_{A_0} W_2\,\mathrm{d}A + \frac{\Omega^2}{2}\oint_{\partial A i}\frac{\partial W_0}{\partial x}\,\mathrm{d}y\right] + \dots \tag{5.1.8}$$

Um komplizierte Integrationen zu vermeiden, wäre es wünschenswert, wenn man die beiden Flächenintegrale entweder auf Linienintegrale längs der Berandung oder auf Flächenintegrale mit Integranden niedriger Ordnung reduzieren könnte. Das gelingt mit Hilfe des GREENschen Satzes

$$\iint_{A_0} (W\,\Delta\lambda - \lambda\,\Delta W)\,\mathrm{d}A = \oint_{\partial A_0}\left(W\frac{\partial\lambda}{\partial n} - \lambda\frac{\partial W}{\partial n}\right)\mathrm{d}s\,. \tag{5.1.9}$$

Setzt man hier $\lambda = \dfrac{r^2}{4}$, so ist $\Delta\lambda = 1$ und das erste Integral liefert den gewünschten Fluß. Das zweite Integral kann mit der LAPLACE- oder Wärmeleitungsgleichung (der das jeweilige w genügt) umgeformt werden. Wir führen diese Methode etwas ausführlicher für den Fluß nullter Ordnung vor. Es gilt

$$\iint_{A_0} W_0\,\mathrm{d}A = \frac{1}{4}\iint_{A_0} r^2\,\Delta W_0\,\mathrm{d}A + \frac{1}{4}\oint_{\partial A_0}\left(W_0\frac{\partial}{\partial n}r^2 - r^2\frac{\partial W_0}{\partial n}\right)\mathrm{d}s\,. \tag{5.1.10}$$

Nun erfüllt W_0 die inhomogene LAPLACEgleichung $\Delta W_0 = -\dfrac{C}{\mu}$ und die Randbedingung $W_0(\partial A_0) = 0$, also ist:

$$Q_0 = \iint_{A_0} W_0\,\mathrm{d}A = -\frac{C}{4\mu}\iint_{A_0} r^2\,\mathrm{d}A - \frac{1}{4}\oint_{\partial A_0} r^2\frac{\partial W_0}{\partial n}\,\mathrm{d}s\,. \tag{5.1.11}$$

Zur Vereinfachung des Flusses zweiter Ordnung benutzen wir $\left(\dfrac{\partial}{\partial t} - \nu\Delta\right) W_2 = -(\boldsymbol{q}_1\cdot\nabla)\,W_1 - (\boldsymbol{q}_2\cdot\nabla)\,W_0$ und erhalten zunächst

$$\iint_{A_0} W_2\,\mathrm{d}A = \frac{1}{4\nu}\iint_{A_0} r^2\left(\frac{\partial}{\partial t}W_2 + (\boldsymbol{q}_1\cdot\nabla)\,W_1 + (\boldsymbol{q}_2\cdot\nabla)\,W_0\right)\mathrm{d}A + \frac{1}{4}\oint_{\partial A_0}\left(W_2\frac{\partial}{\partial n}r^2 - r^2\frac{\partial}{\partial n}W_2\right)\mathrm{d}s\,. \tag{5.1.12}$$

Man hat anscheinend nicht viel gewonnen, weil auf der rechten Seite ein mindestens ebenso unangenehmes Flächenintegral auftritt. Integriert man jedoch über eine zeitliche Periode, so verschwindet das erste Flächenintegral. Aufgrund der Rechts-Links-Symmetrie muß ψ_2 antisymmetrisch und W_0 symmetrisch bezüglich $\varkappa$ sein. Dann ist aber $(\boldsymbol{q}_2\cdot\nabla)\,W_0 = \boldsymbol{K}\cdot(\nabla\psi_2\times\nabla W_0)$ ebenfalls antisymmetrisch und das dritte Flächenintegral verschwindet. Das mittlere Flächenintegral ist zwar ungleich Null, es enthält aber nur noch die einfacher zu berechnenden linearen Terme. Mit der Abkürzung $[\dots]h$ für die zeitliche Mittelung erhält man schließlich:

$$Q = Q_0 + \frac{\varepsilon^2}{4}\left[2\Omega^2\oint_{\partial A_i}\frac{\partial W_0}{\partial x}\,\mathrm{d}y + \oint_{\partial A_0}\left(W_2\frac{\partial}{\partial n}r^2 - r^2\frac{\partial}{\partial n}W_2\right)\mathrm{d}s + \frac{1}{\nu}\iint_{A_0} r^2(\nabla\psi_1\times\nabla W_1)\cdot\mathrm{d}\boldsymbol{A}\right]_T. \tag{5.1.13}$$

Ersichtlich ist das letzte Flächenintegral mathematisch am unangenehmsten. Beschränkt man sich (was wir in dieser Arbeit nicht können) auf Strömungen im engen Spalt zwischen zwei beliebig geformten Zylindern, so daß $[\Delta r \ll r_1]$ gilt, so macht man sicher keinen großen Fehler, wenn man r durch $(\frac{1}{2}(r_1 + r_2))^2$ ersetzt. Das letzte Integral kann unter dieser Voraussetzung näherungsweise ersetzt werden durch

$$\iint_{A_0} r^2(\nabla\psi_1 * \nabla W_1)\,\mathrm{d}\boldsymbol{A} \approx \frac{1}{4}(r_1 + r_2)^2\oint_{\partial A_0}\psi_1\frac{\partial W_1}{\partial s}\,\mathrm{d}s\,. \tag{5.1.14}$$

5.2 Berücksichtigung der speziellen Geometrie

Die Untersuchung in 5.1 galt für beliebige (symmetrische) Berandung und beliebige (periodische) Schwingung des Innenzylinders. Wir spezialisieren jetzt auf zylindrische Symmetrie der Berandung und auf harmonische Schwingungen $\Omega(t) = r_1 \sin \omega t$. Mit $W_0(r)$ aus (3.3.3) berechnet man aus (5.1.11) leicht den Durchfluß nullter Ordnung

$$Q_0 = \frac{\pi C}{8\mu}\left[(r_2^4 - r_1^4) - \frac{(r_2^2-r_1^2)^2}{\ln r_2/r_1}\right]. \tag{5.2.1}$$

Bevor wir den schwierigen frequenzabhängigen Anteil gesondert behandeln, wird noch der frequenzabhängige Teil Q_2^* ausgerechnet:

$$Q_2^* = \frac{\varepsilon^2}{2}[\Omega^2]_T \oint\limits_{\partial A_i} \frac{\partial W_0}{\partial x} dy = \frac{\pi}{4}\varepsilon^2 r_1^3 W_0'(r_1), \qquad W_0'(r_1) = \frac{C}{4\mu r_1}\left[\frac{r_2^2 - r_1^2}{\ln \dfrac{r_2}{r_1}} - 2r_1^2\right]. \tag{5.2.2}$$

Es wird darauf hingewiesen, daß Q_2^* noch nicht den Grenzwert $\lim\limits_{\omega \to 0} Q_2$ symbolisiert! Es bleibt noch übrig, den folgenden Ausdruck auszuwerten:

$$\tilde{Q}_2 = \varepsilon^2\left[\iint\limits_{A_0} W_2 \, dA\right]_T = \varepsilon^2 \iint\limits_{A_0} u \, dA . \tag{5.2.3}$$

Entweder durch direkte Integration des Ausdrucks

$$\tilde{Q}_2 = \frac{2\pi\varepsilon^2}{\ln r_2/r_1} \int\limits_{r_1}^{r_2} [U_p(r_2) \ln r_1/\varrho + U_p(\varrho) \ln r_2/r_1 + Z^*(r_1) \ln \varrho/r_2] \, \varrho \, d\varrho \tag{5.2.4}$$

oder mit Hilfe von (5.1.13) gelingt nach elementarer Umformung die Reduktion auf ein Integral, das nur noch Terme der linearen Theorie enthält:

$$\tilde{Q}_2 = \frac{\pi}{8}\frac{\varepsilon^2}{\nu}\mathrm{Re}\left\{2\int\limits_{r_1}^{r_2} \bar{F}G\varrho \, d\varrho - \frac{r_2^2 - r_1^2}{\ln r_2/r_1}\int\limits_{r_1}^{r_2} \bar{F}G\frac{d\varrho}{\varrho} + \frac{\nu}{2}r_1^2\left(2r_1^2 - \frac{r_2^2 - r_1^2}{\ln r_2/r_1}\right)\left[2i\left(G' + \frac{1}{r}G\right)\Big|_{r_1} - \frac{C}{\mu}\right]\right\}. \tag{5.2.4}$$

Da $\bar{F}(r)$ und $G(r)$ nach 4.2—3 bekannt sind, ist $\tilde{Q}_2$ berechenbar. Es stellt sich die Frage, ob man noch weitere Reduktionen vornehmen könnte, wenn man G wieder durch F ausdrückt und für F die Lösung des linearen Querströmungsproblems einsetzt. Da jedoch bereits die Integration von G auf LOMMELfunktionen führte, ist eine einfache Lösung, die bekannte transzendente Funktionen enthielte, nicht zu erwarten. Dies scheint auch noch aus einem anderen Grund unmöglich zu sein: die komplexe Integrationstheorie der BESSELfunktionen gilt nur für holomorphe Funktionen, aber nicht für Ausdrücke der Gestalt $\int F(\bar{z})\, G(z)\, dz$, deren Ergebnis wegabhängig sein muß. Man ist wohl oder übel gezwungen, (5.2.5) als Endergebnis anzusehen, oder man muß Integrationstheorien entwickeln, die über die LOMMEL- und STRUVEfunktionen hinausgehen. Dies kann nicht Aufgabe dieser Arbeit sein.

Es bleibt noch übrig, dieses Ergebnis numerisch auszuwerten oder im Rahmen einer weiteren Störungsrechnung zu vereinfachen.

5.3 Die Leistung des schwingenden Zylinders

Die Gesamtkraft (pro Längeneinheit), die die Strömung auf den Zylinder ausübt, resultiert aus dem Integral des Druckes und der Schubspannung über die Oberfläche [3]:

$$\boldsymbol{K} = -\iint\limits_{\partial A_i} P\boldsymbol{n} \, dA + \iint\limits_{\partial A_i} T \cdot \boldsymbol{n} \, dA . \tag{5.3.1}$$

Für ein inkompressibles NEWTONsches Fluid lautet der Tensor der Reibungsspannung

$$T = \mu(\nabla\boldsymbol{v} + \nabla\boldsymbol{v}^t) . \tag{5.3.2}$$

Aus der Haftbedingung an der Wand $\boldsymbol{v}(\partial A_i) = \varepsilon\dot{\Omega}(t)\, \boldsymbol{i}$ (man beachte, daß Ω nur von der Zeit, aber nicht vom Ort abhängt!) folgt eine bemerkenswerte Vereinfachung des Schubspannungsintegrals. Hierzu sei ∂s ein beliebiger geschlossener Weg auf dem schwingenden Innenzylinder mit der eingeschlossenen Fläche S. Aus dem (modifizierten) STOKESschen Satz

$$\begin{aligned} 0 &= -2\varepsilon\dot{\Omega}\boldsymbol{i} \times \oint\limits_{\partial S} d\boldsymbol{x} = 2\oint d\boldsymbol{x} \times \boldsymbol{v} = 2\iint\limits_{S} (d\boldsymbol{A} \times \nabla) \times \boldsymbol{v} = \\ &= \iint\limits_{S} [(\nabla\boldsymbol{v} + \nabla\boldsymbol{v}^t)\cdot\boldsymbol{n} - (\nabla \times \boldsymbol{v}) \times \boldsymbol{n} - 2(\nabla \cdot \boldsymbol{v})\boldsymbol{n}] \, dA \end{aligned} \tag{5.3.3}$$

und der vorausgesetzten Inkompressibilität resultiert die vereinfachte Darstellung des Kraftintegrals

$$\boldsymbol{K} = -\iint\limits_{\partial A_i} P\boldsymbol{n} \, dA + \mu \iint\limits_{\partial A_i} (\nabla \times \boldsymbol{v}) \times \boldsymbol{n} \, dA . \tag{5.3.4}$$

Der Wirbelvektor wird durch die Längsströmung und die Querströmungsstromfunktion ausgedrückt:

$$\nabla \times \boldsymbol{v} = \nabla \times (\boldsymbol{k} \times \nabla\psi + w\boldsymbol{K}) = \boldsymbol{k}\,\Delta\psi + \nabla W \times \boldsymbol{k}\,. \tag{5.3.5}$$

Führt man noch die belanglose Integration in z-Richtung aus, so erhält man (mit $P = -Cz + \tilde{P}$) die Ausgangsform

$$\frac{\boldsymbol{K}}{l} = \int\limits_{s(t)} \left(-\tilde{P}\boldsymbol{n} + \mu\,\Delta\psi(\boldsymbol{k} \times \boldsymbol{n}) + \mu\frac{\partial W}{\partial n}\boldsymbol{k}\right) \mathrm{d}s \tag{5.3.6}$$

als Integral über eine beliebige Schnittkurve $s(t)$ der Ebene $z = \text{const.}$ mit dem schwingenden Zylinder (zunächst: beliebigen Querschnitts). Für unsere Untersuchung ist die mittlere Leistung wichtiger als die Gesamtkraft,

$$L = -\frac{\varepsilon}{T}\int\limits_0^T \Omega \cdot \mathrm{d}t \oint\limits_{s(t)} (-\tilde{P}\,\mathrm{d}y + \mu\,\Delta\psi\,\mathrm{d}x)\,. \tag{5.3.7}$$

Hier stört uns noch der unbekannte Druck P. Eine partielle Integration überführt

$$-\oint\limits_{s(t)} \tilde{P}\,\mathrm{d}y = \oint\limits_{s(t)} y\,\mathrm{d}\tilde{P} = \oint\limits_{s(t)} y\,\nabla\tilde{P} \cdot \mathrm{d}\boldsymbol{x}\,. \tag{5.3.8}$$

Aus den Navier-Stokes-Gleichungen (man beachte die Vereinbarung (2.1.5))

$$\operatorname{grad}\tilde{P} + \varrho\frac{D\boldsymbol{q}}{Dt} = \mu\,\Delta\boldsymbol{q}\,; \qquad \boldsymbol{q} = \boldsymbol{k} \times \nabla\psi \tag{5.3.9}$$

kann der Gradient des Druckes durch ψ ausgedrückt werden. Das Beschleunigungsglied brauchen wir im Folgenden nicht zu berücksichtigen, denn auf der Oberfläche des Zylinders ist

$$\frac{\mathrm{D}\boldsymbol{q}}{\mathrm{D}t} = \varepsilon\Omega^{\cdot\cdot}\boldsymbol{i}\,. \tag{5.3.10}$$

Weil wir jedoch über eine Periode integrieren, muß das Integral

$$\int\limits_0^T \Omega^{\cdot}\,\Omega^{\cdot\cdot}\,\mathrm{d}t \oint\limits_{s(t)} y\,\mathrm{d}x = 0 \tag{5.3.11}$$

verschwinden. Wir erhalten schließlich das Leistungsintegral:

$$L = -\frac{\varepsilon\mu}{T}\int\limits_0^T \Omega^{\cdot}\,\mathrm{d}t \left[\oint\limits_{s(t)} \left(1 - y\frac{\partial}{\partial y}\right)\Delta\psi\,\mathrm{d}x + \oint\limits_{s(t)} \left(y\frac{\partial}{\partial x}\right)\Delta\psi\,\mathrm{d}y\right]. \tag{5.3.12}$$

Um im Rahmen der Störungsrechnung konsistent zu bleiben, wird genau wie in 5.1 sowohl der Integrand als auch die zeitabhängige Integrationsgrenze des inneren Integrals nach ε entwickelt und nach Potenzen des Störparameters geordnet. Dies bedingt in (5.3.12) zwei Änderungen. Zunächst wird an Stelle der Integration über den zeitabhängigen Zylinder $s(t)$ nur noch über die starre Ausgangslage s_0 integriert, und weiterhin tritt an Stelle des Ausdrucks jetzt die Entwicklung

$$\Delta\psi \to \varepsilon\,\Delta\psi_1 + \varepsilon^2\left(\Delta\psi_2 + \Omega\frac{\partial}{\partial x}\Delta\psi_1\right) + \dots \tag{5.3.13}$$

Die Leistung darf vom Vorzeichen von ε nicht abhängen (bei vorausgesetzter Rechts-Links-Symmetrie), deshalb können nur die ungeraden Glieder dieser Entwicklung von Null verschiedene Terme ergeben. Andererseits hatten wir die Lösung der linearen Querströmung

$$\left(\frac{\partial}{\partial t} - \nu\Delta\right)\Delta\psi_1 = 0 \Rightarrow \psi_1 = \psi_{11} + \psi_{12} \quad \text{mit} \quad \Delta\psi_{11} = 0; \qquad \left(\frac{\partial}{\partial t} - \nu\Delta\right)\psi_{12} = 0 \tag{5.3.14}$$

in zwei Teile zerlegt, deren erstes Glied ψ_{11} ersichtlich keinen Einfluß auf die mittlere Leistung haben kann. Es bleibt also allein übrig

$$L = \frac{-\varepsilon^2\varrho}{T}\int\limits_0^T \Omega^{\cdot}\,\mathrm{d}t\,\frac{\mathrm{d}}{\mathrm{d}t}\left[\oint\limits_{S_0} \left(1 - y\frac{\partial}{\partial y}\right)\psi_{12}\,\mathrm{d}x + \oint\limits_{S_0} \left(y\frac{\partial}{\partial x}\right)\psi_{12}\,\mathrm{d}y\right] + O(\varepsilon^4)\,. \tag{5.3.15}$$

Für den harmonisch schwingenden Kreiszylinder kann man dieses Integral elementar auswerten:

$$L = -\pi r_1\varepsilon^2\varrho\omega^2 \operatorname{Re}\{iC_2\} + O(\varepsilon^4)\,. \tag{5.3.16}$$

Die Konstante C_2 wurde in (4.2.9) ausgerechnet.

6. Diskussion der Grenzfälle

Eine numerische Berechnung des Durchflusses zweiter Ordnung ist recht aufwendig, weil man zunächst ein Programm zur Berechnung der Besselfunktionen komplexen Arguments aufstellen müßte. Interessiert man sich jedoch für kleine oder große Werte des dimensionslosen Parameters

$$\varepsilon_1 = \sqrt{\frac{\omega r_1^2}{\nu}}, \tag{6.1}$$

so kann man diese Schwierigkeit umgehen, indem man die Zylinderfunktionen

$$Z_n\left(\sqrt{i}\,\varepsilon_1 \frac{r}{r_1}\right) \tag{6.2}$$

als asymptotische Reihen für kleine ε_1 (langsame Schwingungen einer zähen Flüssigkeit) oder große ε_1 (schnelle Schwingungen einer wenig zähen Substanz) darstellt.

Wir behandeln zunächst den einfachen Fall $\varepsilon_1 \ll 1$. Für betragsmäßig kleine Argumente lauten die Reihenentwicklung (siehe Abramowitz [1])

$$I_1(z) = \sum_{k=0}^{\infty} \frac{(z/2)^{2k+1}}{k!(k+1)!}, \tag{6.3}$$

$$K_1(z) = \frac{1}{2} \ln z/2 \cdot I_1(z) - \frac{1}{2} \sum_{k=0}^{\infty} (\Psi(k+1) + \Psi(k+2)) \frac{(z/2)^{2k+1}}{k!(k+1)!}$$

$$\left(\text{mit} \quad \Psi(z) = \Gamma'(z)/\Gamma(z)\right).$$

Für den Durchfluß zweiter Ordnung folgt daraus nach längerer Rechnung und elementarer Integration ($\xi = r_2^2/r_1^2$, $\ln \xi = \tau$)

$$Q_2 = Q_2^* + \tilde{Q}_2 = \frac{\pi}{8} \frac{C}{\mu} \varepsilon^2 r_1^4 \left\{ 2\left(\frac{\xi}{\xi - 1} - \frac{1}{\tau}\right)\left(\frac{\xi - 1}{\tau} - 1\right) + \frac{\varepsilon_1^4}{24} [\eta_1\tau^5 + \eta_2\tau^4 + \eta_3\tau^3 + \eta_4\tau^2 + \eta_6] / \right.$$
$$\left. /[(\xi - 1)^2 \tau^2(\tau(\xi + 1) - 2(\xi - 1))^2] + \dots \right\}, \tag{6.4}$$

$$\eta_1 = \tfrac{3}{2} \xi^3(\xi + 1), \qquad \eta_2 = \tfrac{1}{4} (\xi - 1)(18\xi^4 + 34\xi^3 + 14\xi^2 - \xi - 1),$$
$$\eta_3 = -\tfrac{1}{24} (\xi - 1)^2 (34\xi^4 + 153\xi^3 + 13\xi^2 - 33\xi - 69),$$
$$\eta_4 = -\tfrac{1}{360} (\xi - 1)^3 (134\xi^4 - 711\xi^3 + 1094\xi^2 + 5169\xi + 4574),$$
$$\eta_5 = \tfrac{1}{36} (\xi - 1)^4 (125\xi^3 + 886\xi^2 + 1615\xi + 1130), \; \eta_6 = -\tfrac{1}{12} (\xi - 1)^5 (149\xi^2 + 536\xi + 419).$$

6.1 Der Grenzwert beliebig langsamer Schwingung

Interessant ist der Grenzwert $w \to 0$, das heißt $\varepsilon_1 \to 0$ in (6.4). Hier müßte sich derselbe Durchfluß (bis zu Gliedern der Ordnung ε^2) ergeben wie beim ruhenden exzentrischen Innenzylinder:

$$\lim_{\omega \to 0} Q_2 = \frac{\pi}{4} \frac{C}{\mu} \varepsilon^2 r_1^4 \left(\frac{r_2^2}{r_2^2 - r_1^2} - \frac{1}{2 \ln r_2/r_1}\right)\left(\frac{r_2^2 - r_1^2}{2 r_1^2 \ln r_2/r_1} - 1\right) + \dots \tag{6.1.1}$$

Die Torsion eines zylindrischen Stabes führt mathematisch auf dasselbe Randwertproblem wie die stationäre Längsströmung eines Newtonschen Fluids durch einen Zylinder gleicher Berandung. Schon im Jahre 1893 konnte McDonald [9] das Torsionsmoment eines exzentrischen Kreisringprofils explizit angeben. Diese Lösung wurde von Cardwell (und anderen) benutzt, um den Durchfluß im exzentrischen Ringspalt zu ermitteln. Hier die explizite Lösung in der Nomenklatur der Originalarbeit [12]:

$$Q = \frac{\pi}{8\mu}\left(-\frac{\partial P}{\partial z}\right) r_2^4 \left\{1 - \delta^4 + Z - 4e^2(1 - \delta)^2\left[\delta^2 + \frac{\delta^4}{(1 - e^2(1 - \delta)^2)^2} + \dots\right]\right\},$$

$$\delta = r_1/r_2; \qquad e = \frac{e_1}{r_2 - r_1} \quad (e_1\text{: absolute Exzentrizität}), \tag{6.1.2}$$

$$Z = \frac{4e^2(1 - \delta)^2 S_2^2}{\operatorname{arsinh} S_2 - \operatorname{arsinh} S_2/\delta}; \qquad S_2 = \frac{1 - \delta}{2e} \sqrt{\left[e^2 - \left(\frac{1 + \delta}{1 - \delta}\right)^2\right][e^2 - 1]}.$$

Wir wollen noch nachweisen, daß das Ergebnis unserer Störungsrechnung korrekt mit der Reihenentwicklung der exakten Lösung von Cardwell [5] (bis zu Gliedern zweiter Ordnung) übereinstimmt. Hierzu wird (6.1.2) in eine Taylorreihe bezüglich der absoluten Exzentrizität e_1 an der Stelle $e_1 = 0$ entwickelt. Man berechnet zunächst ($e = e_1/(r_2 - r_1)$)

$$\operatorname{arsinh} S_2 - \operatorname{arsinh} S_2/\delta, \tag{6.1.3}$$

und daraus folgt

$$Z = \frac{4e^2(1 - \delta)^2 S_2^2}{\operatorname{arsinh} S_2 - \operatorname{arsinh} S_2/\delta} = \frac{(1 - \delta)^4}{\ln \delta}\left[\left(\frac{1 + \delta}{1 - \delta}\right)^2 - 2e^2 \frac{(1 + \delta^2)}{(1 - \delta)^2} - \frac{(1 + \delta)}{(1 - \delta)} \frac{e^2}{\ln \delta}\right] + O(e^4). \tag{6.1.4}$$

Setzt man diese Entwicklung in die Gesamtdurchflußmenge, so erhält man

$$Q = \frac{\pi}{8\mu}\left(-\frac{\partial P}{\partial z}\right) r_2^4 \left\{1 - \delta^4 + \frac{(1-\delta^2)}{\ln\delta} - \frac{e_1^2}{r_2^2}\left[\frac{2(1+\delta^2)}{\ln\delta} + \frac{(1-\delta^2)}{(\ln\delta)^2} + \frac{4\delta^2}{(1-\delta^2)}\right]\right\} + O(e_1^4)\,. \tag{6.15}$$

Substituiert man hier $e_1 = \lambda \sin \omega t$ und integriert diesen Durchfluß über eine volle Periode, so erhält man gerade (6.1.1).

6.2 Der Grenzfall schneller Schwingungen

Schwieriger zu behandeln ist der Grenzfall großer Frequenzen $\varepsilon_1 \gg 1$, denn die asymptotische Näherung

$$I_1(z) \approx \frac{1}{\sqrt{2\pi z}}\,e^z\,, \qquad K_1(z) \approx \sqrt{\frac{\pi}{2z}}\,e^{-z}\,. \tag{6.2.1}$$

führt auch hier wieder auf nicht elementar integrierbare komplexe Integrale (Typus: Fehlerintegral), die numerisch ausgewertet wurden. Die Frequenzen dürfen andererseits nicht beliebig groß gewählt werden, denn in der zu Grunde liegenden Störungsrechnung wurden die Störgeschwindigkeiten als klein gegenüber der maximalen Grundgeschwindigkeit angesehen. Die Gültigkeit dieser asymptotischen Näherung liegt also im Intervall

$$1 \ll \varepsilon_1 \ll \frac{r_1}{2\nu}\sqrt{\frac{r_1 C y}{\varrho\varepsilon}} \quad \text{mit} \quad Y = 1 - R^2 + R^2 \ln R^2; \qquad R^2 = \frac{r_2^2 - r_1^2}{2r_1^2 \ln r_2/r_1}\,. \tag{6.2.2}$$

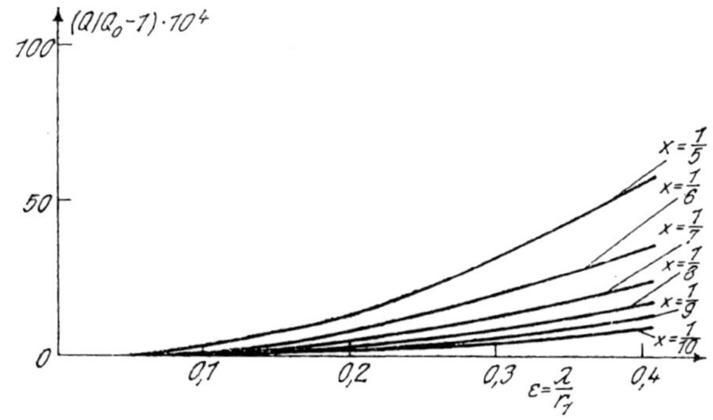

Bild 3. Erhöhung der Durchflußmenge für langsame Schwingungen bei gleichbleibendem Druckgradienten $\varepsilon_1 = \sqrt{\frac{\omega r_1^2}{\nu}} = 0{,}1$; $x = \frac{r_1}{r_2}$

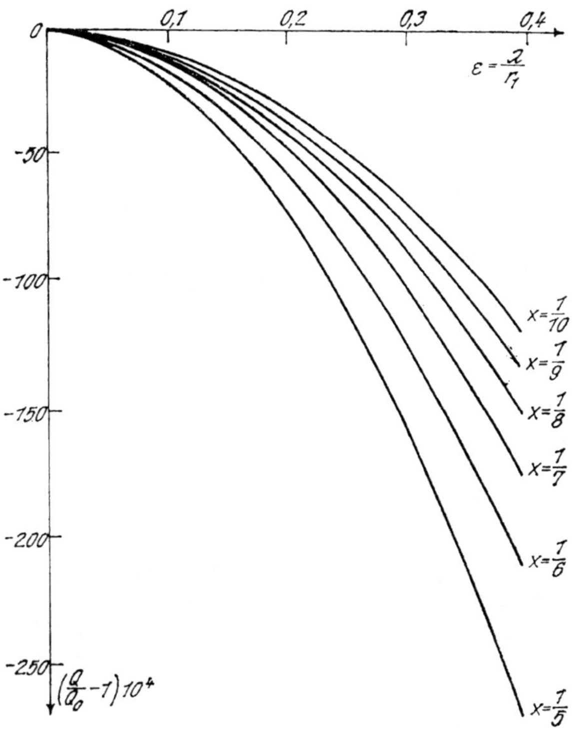

Bild 4. Minderung der Durchflußmenge für schnelle Schwingungen bei gleichbleibendem Druckgradienten $\varepsilon_1 = \sqrt{\frac{\omega r_1^2}{\nu}} = 100$, $x = \frac{r_1}{r_2}$

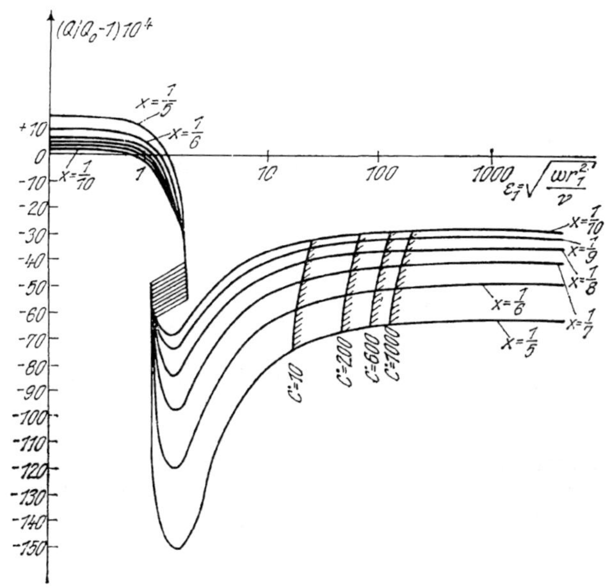

Bild 5. Abhängigkeit der Durchflußmenge von der Frequenz (Grenzfälle) $\varepsilon = \frac{\lambda}{r_1} = 0{,}2$; $C^* = \frac{-\frac{\partial P}{\partial z}}{(\varrho\nu^2/r_1^3)}$

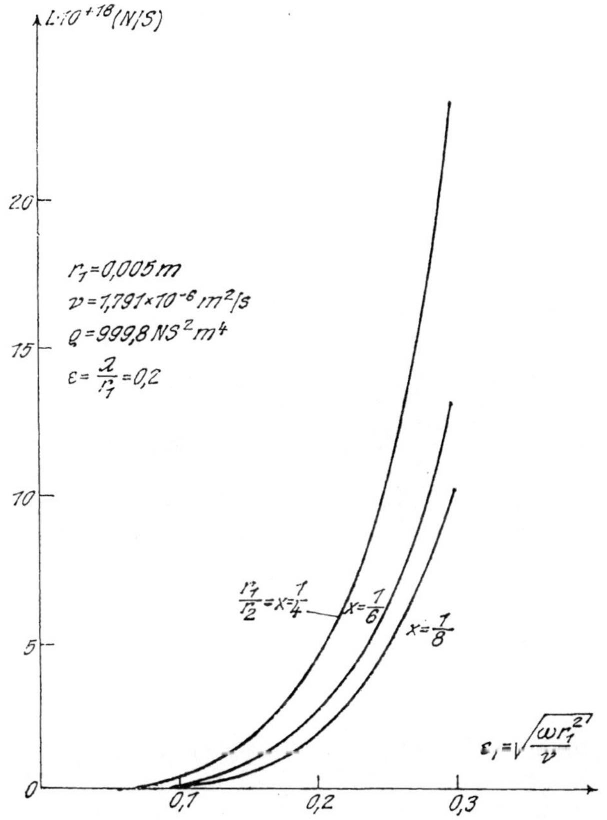

Bild 6. Erhöhung der Leistung des Innenzylinders für langsame Schwingungen bei wachsender Frequenz

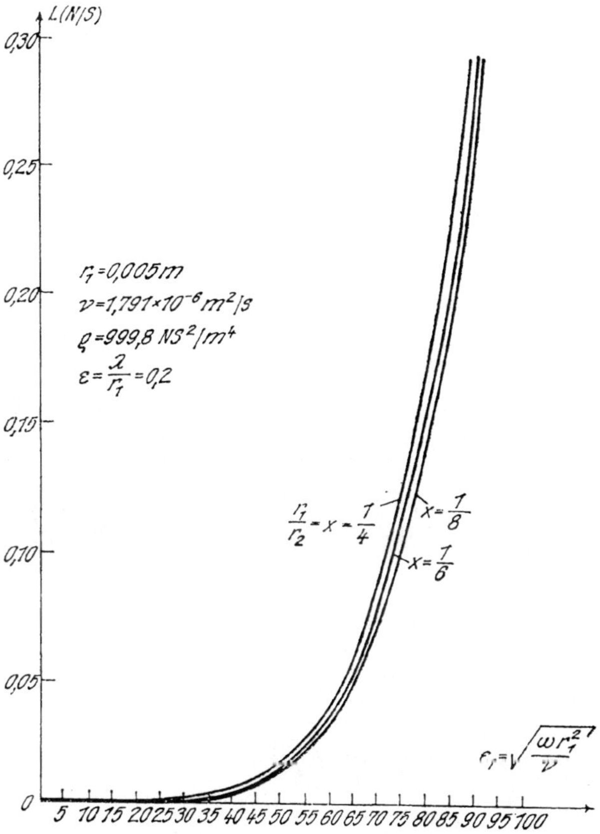

Bild 7. Erhöhung der Leistung des Innenzylinders für schnelle Schwingungen bei wachsender Frequenz

Literatur

1 ABRAMOWITZ, M., STEGUN, I. A., Handbook of mathematical functions with formulas, graphs and mathematical tables, Dover Public. Inc., New York.
2 AMES, W. F., Nonlinear partial differential equations in engineering, Academic Press, New York, 1965.
3 BECKER, E., Gasdynamik, Teubner, Stuttgart 1966.
4 BELLMANN, R., Methoden der Störungsrechnung in Mathematik, Physik und Technik; R. Oldenburg Verlag, München-Wien, 1967.
5 CARDWELL, F. W., ROY, J., Techn. Coll.,Glasgow 1930.
6 HOLTSMARK, J., JOHNSEN, I., SIKKELAND, T. and SKAVLEM, S., Boundary layer flow near a cylindrical obstacle in oscillating incompressible fluid, J. of the Acoustical Society America v. 26.
7 KREYSZIG, Advanced Engineering Methods, Wiley.
8 LUKE, Y. L., Integrals of BESSEL Functions, McGraw-Hill, Book Company, INC. 1962.
9 MCDONALD, H. M., On the torsional strength of a hollow shaft, Proc. Cambrige Phil. Soc. 8 (1894) 62.
10 MCLACHLAN, BESSEL functions for engineers, Oxford 1954.
11 SCHLICHTING, H., Grenzschichttheorie, Verlag G. BRAUN, Karlsruhe 1958.
12 TIEDT, W., Berechnung des laminaren und turbulenten Reibungswiderstandes konzentrischer und exzentrischer Ringspalte, TH Darmstadt, März 1968.

Eingereicht am 30. 10. 1980

Anschrift: Dc. MOBARAK N. FARAH, Egypt, El-Mansoura, El-Mansoura University, Faculty of Scie., Dept. of Maths.

ZAMM 61, 379–393 (1981)

M. TASCHE

Eine einheitliche Herleitung verschiedener Interpolationsformeln mittels der Taylorschen Formel der Operatorenrechnung

Zahlreiche Verfahren der numerischen Differentiation und Integration basieren auf Interpolationsmethoden oder Polynomentwicklungen. Deshalb ist eine einheitliche Herleitung von Interpolationsformeln und Polynomentwicklungen und entsprechenden Fehlerabschätzungen von Interesse. In der vorliegenden Arbeit lösen wir dieses Problem mittels der Taylorschen Formel der Operatorenrechnung. Dabei wird die Operatorenrechnung als Teilgebiet der Theorie der verallgemeinert inversen Operatoren verstanden.

Numerous methods of numerical differentiation and integration are based on methods of interpolation or polynomial expansions. Therefore a unified approach of interpolation formulas and polynomial expansions and appropriate error estimations is interesting. In the present paper we solve this problem by means of Taylor's formula of operational calculus. Here we understand the operational calculus as a part of the theory of generalized inverse operators.

Многочисленные методы численного дифференцирования и интегрирования основываются на способах интерполяции или полиномиальных разложениях. Поэтому единый вывод интерполяционных формул и полиномиальных разложений и соответствующих оценок погрешности представляет интерес. В данной работе эта проблема решена посредством формулой Тэйлора операторного исчисления. При этом операторное исчисление понимается как часть теории обобщенно-обратных операторов.

Es sei $\mathfrak{B}$ ein komplexer Vektorraum mit dem Nullvektor o. Wir betrachten im folgenden nur *lineare* Operatoren aus $\mathfrak{B}$ in sich. Der Einheitsoperator sei mit I und der Nulloperator mit O bezeichnet. Für einen gegebenen Operator A aus $\mathfrak{B}$ in sich sei $\mathfrak{D}(A)$ dessen Definitionsbereich, $\mathfrak{R}(A)$ dessen Wertebereich und $\mathfrak{N}(A)$ dessen Kern. Wir verwenden dann die Schreibweise $A: \mathfrak{D}(A) \to \mathfrak{B}$. Für einen linearen Teilraum $\mathfrak{M}$ mit $\mathfrak{M} \subseteqq \mathfrak{D}(A)$ sei $A \mid \mathfrak{M}$ die Einschränkung von A auf $\mathfrak{M}$. Ferner wird $A^0 = I$ vereinbart.

Es sei ein Operator $D: \mathfrak{D} \to \mathfrak{B}$ mit $\mathfrak{D} = \mathfrak{D}(D) \subseteqq \mathfrak{B}$ gegeben, der eine *äußere Inverse* $T: \mathfrak{B} \to \mathfrak{B}$ besitzt, d. h. es gilt nach Definition ([6]) $\mathfrak{R}(T) \subseteqq \mathfrak{D}$ und $TDT = T$. Bildet man mittels D und T den Operator $F = I - TD$: $\mathfrak{D} \to \mathfrak{B}$, so ist F wegen

$$F^2 = I - TD - TD + (TDT)\, D = F$$

auf $\mathfrak{D}(D)$ ein Projektor. Dann gilt für alle $x \in \mathfrak{D}(D^p)$, $p = 1, 2, \ldots$, die sogenannte TAYLORsche *Formel der Operatorenrechnung*

$$x = \sum_{j=0}^{p-1} T^j F D^j x + T^p D^p x\,, \tag{1}$$

wie man leicht nachrechnet. Es sei noch bemerkt, daß die Gleichung (1) auch dann besteht, wenn T keine äußere Inverse von D ist.

Wir erinnern daran, daß $T: \mathfrak{B} \to \mathfrak{B}$ eine *Rechtsinverse* von $D: \mathfrak{D} \to \mathfrak{B}$ ist, falls $\mathfrak{R}(T) \subseteqq \mathfrak{D}$ und $DT = I$ gilt; daß $T: \mathfrak{B} \to \mathfrak{B}$ eine *verallgemeinerte Inverse* von $D: \mathfrak{D} \to \mathfrak{B}$ ist, wenn $\mathfrak{R}(T) \subseteqq \mathfrak{D}$ und $DTD = D$ sowie $TDT = T$ gilt; und daß $T: \mathfrak{B} \to \mathfrak{B}$ eine DRAZIN-*Inverse* von $D: \mathfrak{D} \to \mathfrak{B}$ *mit dem Index* 1 ist, falls $\mathfrak{R}(T) \subseteqq \mathfrak{D}$ und $DT \mid \mathfrak{D} = TD$, $TDT = T$, $D = D^2T \mid \mathfrak{D}$ gilt ([6]). Wie man sieht, ist T in allen drei Fällen eine äußere Inverse von D.

Zur Herleitung eines Interpolationsverfahrens benutzen wir folgende Methode: Wir wählen $\mathfrak{B}$ als einen Funktionenraum, D als Differentialoperator und konstruieren einen passenden Integraloperator T. Auf Grund der TAYLORschen Formel (1) ordnen wir einer Funktion $x(t) \in \mathfrak{D}(D^p)$, $p = 1, 2, \ldots$, die Näherungsfunktion

$$\sum_{j=0}^{p-1} T^j F D^j x(t)$$

mit dem zugehörigen Restglied $T^p D^p x(t)$ zu. Wie wir in den Abschnitten 4.–11. zeigen werden, lassen sich viele bekannte Interpolationsmethoden in dieser einfachen Weise begründen.

Ist $x \in \mathfrak{D}_\infty$ mit

$$\mathfrak{D}_\infty = \bigcap_{p=1}^{\infty} \mathfrak{D}(D^p)\,,$$

$\mathfrak{B}$ ein normierter Raum und T ein beschränkter Operator, so konvergiert die TAYLOR*sche Reihe*

$$\sum_{j=0}^{\infty} T^j F D^j x \tag{2}$$

genau dann (in der Norm von $\mathfrak{B}$) gegen x, wenn

$$\lim_{p\to\infty} ||T^p D^p x|| = 0$$

ist. Wegen $||T^p D^p x|| \leqq ||T^p||\, ||D^p x||$, $p = 1, 2, \ldots$, ist dann

$$\lim_{p\to\infty} ||T^p||\, ||D^p x|| = 0$$

eine hinreichende Bedingung für die Konvergenz der TAYLORschen Reihe (2).

Bei Anwendungen ist der Differentialoperator D unter gewissen Randbedingungen zu betrachten. Die Operatoren T und F sind zunächst unbekannt und müssen aus D und den Randbedingungen bestimmt werden. Aus diesem Grunde führen wir im ersten Teil dieser Arbeit drei Konstruktionsverfahren für die Operatoren T und F an. Dabei ist das erste Verfahren ein Spezialfall von [3] und [8], während die anderen beiden Methoden durch Verallgemeinerung von [5] und [7] erhalten wurden (siehe auch [9]).

Im zweiten Teil der vorliegenden Note erfolgt die Anwendung der drei Konstruktionsverfahren. Zuerst werden die klassischen Interpolationsverfahren von Lagrange, Newton und Hermite hergeleitet. Interessant ist der Zusammenhang zwischen diesen Verfahren und den nachfolgenden Entwicklungssätzen nach Bernoullischen, Eulerschen, Lidstoneschen und Hermiteschen Polynomen, die — mit Ausnahme der letztgenannten — *keine Orthogonalsysteme* bilden und auch *keine Eigenfunktionen* von D sind.

Es sei noch darauf hingewiesen, daß die Entwicklungssätze nach Bernoullischen, Eulerschen und Lidstoneschen Polynomen in enger Beziehung zu einer Methode der Konvergenzbeschleunigung von Fourierschen Reihen stehen ([9]). Ist $\mathfrak{B}$ ein Hilbert-Raum und ist T vollstetig und selbstadjungiert, so kann man die Konvergenz einer Entwicklung eines Elementes $x \in \mathfrak{D}(D^p)$ nach Eigenvektoren von T dadurch verbessern, daß man x als Summe der endlichen Taylorschen Entwicklung

$$\sum_{j=0}^{p-1} T^j F D^j x$$

und der schnell konvergierenden Entwicklung des Restgliedes $T^p D^p x$ nach Eigenvektoren darstellt. Gewissermaßen dual zu diesem Verfahren ist eine Methode zur Konvergenzverbesserung von Taylorschen Reihen, die Ergebnisse von R. C. Buck (siehe [4], S. 15) und J. K. Shaw ([7]) verallgemeinert. Danach läßt sich der Konvergenzbereich einer Taylorschen Reihe durch Abspaltung einer endlichen Entwicklung nach Eigenvektoren erweitern (siehe Sätze 4 und 8).

1. Erste Konstruktionsmethode

Es sei $\mathfrak{B}$ ein komplexer Vektorraum und $D\colon \mathfrak{D} \to \mathfrak{B}$ ein linearer rechtsinvertierbarer Operator mit dem Definitionsbereich $\mathfrak{D} = \mathfrak{D}(D) \subset \mathfrak{B}$. Der Kern $\mathfrak{N} = \mathfrak{N}(D)$ von D sei n-dimensional $(1 \leqq n < \infty)$ und u_j, $j = 1, 2, \dots, n$, sei eine Basis von $\mathfrak{N}$. Ferner sei $T_0\colon \mathfrak{B} \to \mathfrak{B}$ eine lineare Rechtsinverse von D.

Wir betrachten n lineare Funktionale $R_j\colon \mathfrak{D} \to \mathscr{C}$, $j = 1, 2, \dots, n$, die linear unabhängig sind. Dabei ist $\mathscr{C}$ die Menge der komplexen Zahlen. Dann sei $\mathfrak{D}_1 = \{x \in \mathfrak{D};\ R_j x = 0, j = 1, 2, \dots, n\}$ und $D_1 = D \mid \mathfrak{D}_1$ die Einschränkung von D auf $\mathfrak{D}_1$.

Wir setzen voraus, daß *Null kein Eigenwert* von D_1 ist. Folglich ist die komplexe (n, n)-Matrix $\boldsymbol{G} = (R_j u_k)$, $j, k = 1, 2, \dots, n$, invertierbar. Denn andernfalls würde es $\alpha_j \in \mathscr{C}$, $j = 1, 2, \dots, n$, geben, die nicht sämtlich verschwinden und für die

$$D_1(\alpha_1 u_1 + \alpha_2 u_2 + \dots + \alpha_n u_n) = o$$

gilt, was nach Voraussetzung unmöglich ist. Wie man später an Hand der Beispiele in den Abschnitten 4.—6. sieht, ist die Invertierbarkeit von $\boldsymbol{G}$ eine Verallgemeinerung der bekannten Haarschen Bedingung.

Der Operator $F\colon \mathfrak{D} \to \mathfrak{N}$ wird mittels der Gleichung

$$Fx = \sum_{j=1}^{n} \xi_j u_j$$

mit

$$(\xi_1, \xi_2, \dots, \xi_n)^{\mathsf{T}} = \boldsymbol{G}^{-1}(R_1 x, R_2 x, \dots, R_n x)^{\mathsf{T}}$$

für $x \in \mathfrak{D}$ erklärt. Offenbar ist F ein Projektor auf $\mathfrak{N}$, denn für $z \in \mathfrak{N}$ mit

$$z = \zeta_1 u_1 + \zeta_2 u_2 + \dots + \zeta_n u_n$$

gilt

$$R_j z = \zeta_1 R_j u_1 + \zeta_2 R_j u_2 + \dots + \zeta_n R_j u_n\,, \qquad j = 1, 2, \dots, n\,,$$

und somit

$$(R_1 z, R_2 z, \dots, R_n z)^{\mathsf{T}} = \boldsymbol{G}(\zeta_1, \zeta_2, \dots, \zeta_n)^{\mathsf{T}}\,,$$

woraus sich $Fz = z$ und damit $F^2 = F$ auf $\mathfrak{D}$ ergibt. Man nennt F den *Randprojektor* ([8]).

Insbesondere kann man zu einer Basis y_j, $j = 1, 2, \dots, n$, von $\mathfrak{N}$ mit der Eigenschaft $R_j y_k = \delta_{jk}$, $j, k = 1, 2, \dots, n$, übergehen, wobei δ_{jk} das Kronecker-Symbol ist. In diesem Fall lautet der Operator $F\colon \mathfrak{D} \to \mathfrak{N}$

$$Fx = \sum_{j=1}^{n} (R_j x)\, y_j \tag{3}$$

für $x \in \mathfrak{D}$. Mit Hilfe von F bilden wir den Operator $T\colon \mathfrak{B} \to \mathfrak{B}$

$$T = T_0 - F T_0\,, \tag{4}$$

der Green*scher Operator* genannt wird ([8]). Offensichtlich ist T eine Inverse von D_1 und eine Rechtsinverse von D, d. h. $D_1 T = I$, $T D_1 = I \mid \mathfrak{D}_1$ und $DT = I$. Ferner gilt die Gleichung

$$F = I - TD\,. \tag{5}$$

Auf Grund der TAYLORschen Formel (1) erhält man somit:

Satz 1: *Es sei Null kein Eigenwert von D_1 und y_j, $j = 1, 2, \ldots, n$, eine Basis von $\mathfrak{N}$ mit $R_j y_k = \delta_{jk}$, $j, k = 1, 2, \ldots, n$. Dann gilt für jedes $x \in \mathfrak{D}(D^p)$, $p = 1, 2, \ldots$, die TAYLORsche Formel*

$$x = \sum_{k=0}^{p-1} \sum_{j=1}^{n} (R_j D^k x)\, T^k y_j + T^p D^p x\,. \tag{6}$$

2. Zweite Konstruktionsmethode

Die zweite Methode ist ein Spezialfall der ersten. Dazu sei $\mathfrak{B}$ ein separabler komplexer HILBERT-Raum mit dem Skalarprodukt $(.,.)$, $\mathfrak{D}$ sowie $\mathfrak{D}_1$ seien dicht in $\mathfrak{B}$, T_0 sei beschränkt, und D_1 sei ein selbstadjungierter Operator mit reinem Punktspektrum. Wir setzen wiederum voraus, daß *Null kein Eigenwert* von D_1 ist.

Wir nehmen an, daß D_1 genau die Eigenwerte λ_k, $k = 1, 2, \ldots$, endlicher Vielfachheit mit $0 < |\lambda_1| \leqq |\lambda_2| \leqq \ldots$ und $|\lambda_k| \to \infty$ für $k \to \infty$ sowie den zugehörigen orthonormierten Eigenvektoren $x_k \in \mathfrak{D}_1$ besitzt. Das System aller Eigenvektoren von D_1 ist vollständig in $\mathfrak{B}$.

Wie in **1.** kann man die Operatoren F und T konstruieren. Außerdem gilt

$$Tx = \sum_{k=1}^{\infty} \lambda_k^{-1} (x, x_k)\, x_k \tag{7}$$

für $x \in \mathfrak{B}$, so daß T ein vollstetiger selbstadjungierter Operator mit der Norm $||T|| = |\lambda_1|^{-1}$ ist. Offenbar ist $Tx_k = \lambda_k^{-1} x_k$, $k = 1, 2, \ldots$, und somit

$$T^p x = \sum_{k=1}^{\infty} \lambda_k^{-p} (x, x_k)\, x_k\,, \qquad p = 1, 2, \ldots\,, \tag{8}$$

für $x \in \mathfrak{B}$. Folglich ist $||T^p|| = |\lambda_1|^{-p}$. Da für $x \in \mathfrak{D}_\infty$ und $p = 1, 2, \ldots$ dann $||T^p D^p x|| \leqq |\lambda_1|^{-p}\, ||D^p x||$ gilt, folgt aus (6):

Satz 2: *Unter den gemachten Voraussetzungen läßt sich jedes Element $x \in \mathfrak{D}_\infty$ mit der Eigenschaft*

$$\lim_{p\to\infty} |\lambda_1|^{-p}\, ||D^p x|| = 0 \tag{9}$$

in eine konvergente TAYLORsche Reihe entwickeln:

$$x = \sum_{k=0}^{\infty} \sum_{j=1}^{n} (R_j D^k x)\, T^k y_j\,. \tag{10}$$

Wir bemerken, daß für den Eigenvektor x_1 die Bedingung (9) nicht erfüllt ist und daß man x_1 auch nicht in Form einer TAYLORschen Reihe (10) darstellen kann.

Nunmehr soll Satz 2 verallgemeinert werden. Wir bilden für $m \geqq 1$ den beschränkten Operator $T_m\colon \mathfrak{B} \to \mathfrak{B}$ mit

$$T_m x = Tx - \sum_{k=1}^{m} \lambda_k^{-1} (x, x_k)\, x_k = \sum_{k=m+1}^{\infty} \lambda_k^{-1} (x, x_k)\, x_k \tag{11}$$

für $x \in \mathfrak{B}$. Der Operator T_m ist selbstadjungiert und vollstetig und besitzt die Norm $||T_m|| = |\lambda_{m+1}|^{-1}$. Ferner ist $||T_m^p|| = |\lambda_{m+1}|^{-p}$, $p = 1, 2, \ldots$. Nach Konstruktion gilt

$$D_1 T_m x = D_1 T x - \sum_{k=1}^{m} \lambda_k^{-1} (x, x_k)\, D_1 x_k = x - \sum_{k=1}^{m} (x, x_k)\, x_k$$

für $x \in \mathfrak{B}$ und

$$T_m D_1 x = T D_1 x - \sum_{k=1}^{m} \lambda_k^{-1} (D_1 x, x_k)\, x_k = x - \sum_{k=1}^{m} \lambda_k^{-1} (x, D_1 x_k)\, x_k = x - \sum_{k=1}^{m} (x, x_k)\, x_k$$

für $x \in \mathfrak{D}_1$. Hieraus ergeben sich die Gleichungen

$$D_1 T_m x = T_m D_1 x$$

für $x \in \mathfrak{D}_1$ und

$$T_m D_1 T_m x = T_m x$$

für $x \in \mathfrak{B}$. Infolgedessen ist T_m eine äußere Inverse von D_1, aber wegen $D_1 T_m D_1 x \neq D_1 x$ für $x \in \mathfrak{D}_1$ keine verallgemeinerte Inverse von D_1.

Nun führen wir den Operator $F_m\colon \mathfrak{D} \to \mathfrak{B}$ durch $F_m = I - T_m D$ ein. Dann gilt nach (3), (5) und (11) für $x \in \mathfrak{D}$

$$\begin{aligned} F_m x &= x - T_m D x = Fx + \sum_{k=1}^{m} \lambda_k^{-1} (Dx, x_k)\, x_k = \sum_{j=1}^{n} (R_j x)\, y_j + \sum_{k=1}^{m} (TDx, x_k)\, x_k = \\ &= \sum_{j=1}^{n} (R_j x)\, y_j + \sum_{k=1}^{m} (x, x_k)\, x_k - \sum_{k=1}^{m} (Fx, x_k)\, x_k = \\ &= \sum_{k=1}^{m} (x, x_k)\, x_k + \sum_{j=1}^{n} (R_j x) \left[y_j - \sum_{k=1}^{m} (y_j, x_k)\, x_k \right]. \end{aligned} \tag{12}$$

Durch Anwendung der TAYLORschen Formel

$$x = \sum_{k=0}^{p-1} T_m^k F_m D^k x + T_m^p D^p x$$

für $x \in \mathfrak{D}(D^p)$, $p = 1, 2, \ldots$, bekommt man wegen (11) und (12):

Satz 3: *Unter den obigen Voraussetzungen gilt für $x \in \mathfrak{D}(D^p)$, $p = 1, 2, \ldots$, und $m \geqq 1$ die Gleichung*

$$x = F_m x + \sum_{k=1}^{p-1} \sum_{j=1}^{n} (R_j D^k x)\, T_m^k y_j + T_m^p D^p x \tag{13}$$

mit

$$T_m^k y_j = T^k y_j - \sum_{i=1}^{m} \lambda_i^{-k} (y_j, x_i)\, x_i \,. \tag{14}$$

Da für $x \in \mathfrak{D}_\infty$ und $p = 1, 2, \ldots$ dann $\|T_m^p D^p x\| \leqq |\lambda_{m+1}|^{-p} \|D^p x\|$ gilt, ergibt sich aus (13):

Satz 4: *Jedes Element $x \in \mathfrak{D}_\infty$ mit der Eigenschaft*

$$\lim_{p \to \infty} |\lambda_{m+1}|^{-p} \|D^p x\| = 0 \tag{15}$$

für gewisses $m \geqq 1$ läßt sich in eine konvergente TAYLORsche Reihe entwickeln:

$$x = \sum_{k=0}^{\infty} T_m^k F_m D^k x = F_m x + \sum_{k=1}^{\infty} \sum_{j=1}^{n} (R_j D^k x)\, T_m^k y_j. \tag{16}$$

Satz 4 ist eine Verallgemeinerung von Ergebnissen von R. C. BUCK (siehe Abschnitt 9.) und J. K. SHAW ([7]). Als Folgerung von Satz 4 erhält man eine Erweiterung eines Resultates von I. J. SCHOENBERG (siehe Abschnitt 9.):

Folgerung 5: *Jedes Element $x \in \mathfrak{D}_\infty$ mit der Eigenschaft (15) für gewisses $m \geqq 1$ und $R_j D^k x = 0$, $j = 1, 2, \ldots, n$; $k = 0, 1, \ldots$, ist eine Linearkombination der Eigenvektoren $x_1, x_2, \ldots, x_m$.*

3. Dritte Konstruktionsmethode

Die vorher benutzten Bezeichnungen behalten wir bei. Im Unterschied zur zweiten Methode setzen wir jetzt voraus, daß $\lambda_0 = 0$ ein *einfacher Eigenwert* von D_1 mit dem normierten Eigenvektor x_0 ist. Das System aller Eigenvektoren x_k, $k = 0, 1, \ldots$, ist vollständig in $\mathfrak{H}$.

Wir betrachten zuerst den Fall dim $\mathfrak{N} = n > 1$. Für jede beliebige Basis $u_1, u_2, \ldots, u_n$ von $\mathfrak{N}$ ist der Rang der (n, n)-Matrix $(R_j u_k)$, $j, k = 1, 2, \ldots, n$, gleich $n - 1$. Wir können annehmen, daß $u_1 = x_0$ ist und daß die $(n - 1, n - 1)$-Matrix $(R_j u_k)$, $j, k = 2, 3, \ldots, n$, invertierbar ist. Dann existieren eindeutig bestimmte, linear unabhängige Vektoren $q_2, q_3, \ldots, q_n$ aus $\mathfrak{N}$ mit

$$R_j q_k = \delta_{jk} \,, \qquad (q_k, x_0) = 0 \,, \qquad j, k = 2, 3, \ldots, n \,, \tag{17}$$

denn man setzt

$$q_k = \alpha_{2k} u_2 + \alpha_{3k} u_3 + \ldots + \alpha_{nk} u_n \,, \qquad k = 2, 3, \ldots, n \,,$$

mit $\alpha_{jk} \in \mathscr{C}$, $j, k = 2, 3, \ldots, n$, die durch die Gleichungen

$$\delta_{jk} = R_j q_k = \alpha_{2k} R_j u_2 + \alpha_{3k} R_j u_3 + \ldots + \alpha_{nk} R_j u_n \,, \qquad j, k = 2, 3, \ldots, n \,,$$

eindeutig bestimmt sind. Außerdem existiert genau ein Vektor $q_1 \in \mathfrak{D}$ und eine solche komplexe Zahl $\alpha \neq 0$, daß

$$D q_1 = \alpha x_0 \,, \qquad R_k q_1 = \delta_{k1} \,, \qquad k = 1, 2, \ldots, n \,, \qquad (q_1, x_0) = 0 \,, \tag{18}$$

gilt. Denn ist $u \in \mathfrak{D}$ eine spezielle Lösung von $Du = x_0$, so ist

$$q = u + \beta_1 x_0 + \sum_{j=2}^{n} \beta_j q_j$$

mit beliebigen $\beta_j \in \mathscr{C}$, $j = 1, 2, \ldots, n$, die allgemeine Lösung von $Du = x_0$. Wenn q den Bedingungen $R_j q = 0$, $j = 2, 3, \ldots, n$, und $(q, x_0) = 0$ genügen soll, so muß

$$\beta_1 = -(u, x_0) \,, \qquad \beta_j = (u, x_0)\, R_j x_0 - R_j u \,, \qquad j = 2, 3, \ldots, n \,,$$

gelten. Ferner ist $R_1 q \neq 0$, denn andernfalls wäre $q \in \mathfrak{D}_1$ und somit

$$1 = (x_0, x_0) = (x_0, D_1 q) = (D_1 x_0, q) = (o, q) = 0 \,.$$

Setzt man $\alpha = (R_1 q)^{-1}$, so besitzt der Vektor $q_1 = \alpha q$ die gewünschten Eigenschaften (18).

Wir führen die linear unabhängigen Funktionale $U_j \colon \mathfrak{D} \to \mathscr{C}$, $j = 1, 2, \ldots, n$, durch die Gleichungen

$$U_1 = R_1 - (R_1 q_2)\, R_2 - (R_1 q_3)\, R_3 - \ldots - (R_1 q_n)\, R_n \,, \qquad U_j = R_j \,, \qquad j = 2, 3, \ldots, n \,,$$

ein. Dann gilt wegen (17) und (18)

$$U_j q_k = \delta_{jk} \,, \qquad j, k = 1, 2, \ldots, n \,. \tag{19}$$

Im Fall $\dim \mathfrak{N} = n = 1$ bestimmen wir $q_1 \in \mathfrak{D}$ als eindeutige Lösung von

$$Dq_1 = \alpha x_0 , \qquad R_1 q_1 = 1 , \qquad (q_1, x_0) = 0$$

für eine geeignete Zahl $\alpha \neq 0$ und setzen $U_1 = R_1$.

Nun sei $\mathfrak{M}$ die direkte Summe

$$\mathfrak{M} = \mathfrak{N} \oplus \operatorname{span} q_1 ,$$

so daß $\dim \mathfrak{M} = n + 1$ ist. Der Operator $F\colon \mathfrak{D} \to \mathfrak{M}$ sei für $x \in \mathfrak{D}$ durch die Gleichung

$$Fx = (x, x_0)\, x_0 + \sum_{j=1}^{n} (U_j x)\, q_j \tag{20}$$

erklärt. Offenbar ist F ein Projektor auf $\mathfrak{M}$, denn für $y \in \mathfrak{M}$ gilt

$$y = \eta_0 x_0 + \eta_1 q_1 + \dots + \eta_n q_n$$

mit gewissen $\eta_j \in \mathcal{C}$, $j = 0, 1, \dots, n$, so daß sich wegen (17), (18) und (19)

$$\eta_0 = (y, x_0) , \qquad \eta_j = U_j y , \qquad j = 1, 2, \dots, n ,$$

und damit $Fy = y$ ergibt. Hieraus folgt $F^2 = F$ auf $\mathfrak{D}$. Ferner ist $Fz = z$ für $z \in \mathfrak{N}$. Man nennt F den *Randprojektor.*

Jetzt bilden wir den vollstetigen selbstadjungierten Operator $T\colon \mathfrak{B} \to \mathfrak{B}$ der Form (7), der die Norm $\|T\| = |\lambda_1|^{-1}$ besitzt. Offenbar gilt

$$Tx_0 = o , \qquad Tx_k = \lambda_k^{-1} x_k , \qquad k = 1, 2, \dots . \tag{21}$$

Außerdem gilt (8) und $\|T^p\| = |\lambda_1|^{-p}$. Somit erhält man für $x \in \mathfrak{B}$

$$DTx = D_1 Tx = x - (x, x_0)\, x_0 \tag{22}$$

und für $x \in \mathfrak{D}_1$

$$TD_1 x = x - (x, x_0)\, x_0 , \tag{23}$$

so daß $TD_1 = D_1 T \mid \mathfrak{D}_1$ ist. Wegen $D_1 x_0 = Tx_0 = o$ folgen aus (22) und (23) die Gleichungen

$$TD_1 T = T , \qquad D_1 T D_1 = D_1 , \qquad D_1^2 T \mid \mathfrak{D}_1 = D_1 ,$$

d. h., T ist sowohl eine verallgemeinerte Inverse von D_1 als auch eine DRAZIN-Inverse mit dem Index 1 von D_1.

Wir bemerken noch, daß sich aus (22) die Gleichungen

$$TDT = T , \qquad D^2 T \mid \mathfrak{D} = D$$

ergeben. Der Operator T ist also eine äußere Inverse von D. Man nennt T den *GREENschen Operator.*

Ferner gilt auf $\mathfrak{B}$ die Gleichung

$$T = T_0 - FT_0 , \tag{24}$$

welche es bei Anwendungen erlaubt, den Operator T mittels T_0 und F einfach, d. h. ohne Benutzung der Spektraldarstellung (7), zu berechnen. Denn auf Grund der Vollständigkeit des Orthonormalsystems $\{x_k\}$, $k = 0, 1, \dots$, ergibt sich die Behauptung einerseits aus (21) und andererseits aus (20) und

$$T_0 x_0 - FT_0 x_0 = T_0 x_0 - T_0 x_0 = o ,$$

$$T_0 x_k - FT_0 x_k = (\lambda_k^{-1} x_k + z_k) - F(\lambda_k^{-1} x_k + z_k) = \lambda_k^{-1} x_k , \qquad k = 1, 2, \dots ,$$

mit geeignetem $z_k \in \mathfrak{N}$.

Für den Projektor $F\colon \mathfrak{D} \to \mathfrak{M}$ erhält man schließlich die Gleichung

$$F = I - TD , \tag{25}$$

denn wegen (24) gilt für $x \in \mathfrak{D}$

$$x - TDx = x - (T_0 - FT_0)\, Dx = x - T_0 Dx + FT_0 Dx = F(x - T_0 Dx) + FT_0 Dx = Fx ,$$

da $x - T_0 Dx \in \mathfrak{N}$ und damit $F(x - T_0 Dx) = x - T_0 Dx$ ist. Aus (18) und (20) ergibt sich die Gleichung

$$DFx = (U_1 x)\, Dq_1 = \alpha (U_1 x)\, x_0$$

für $x \in \mathfrak{D}$. Andererseits ist wegen (22) und (25)

$$DFx = D(x - TDx) = (I - DT)\, Dx = (Dx, x_0)\, x_0 . \tag{26}$$

Somit erhält man für $x \in \mathfrak{D}$ die Beziehung

$$(Dx, x_0) = \alpha U_1 x .$$

Auf Grund der TAYLORschen Formel (1) erhält man wegen (20):

Satz 6: *Es sei Null ein einfacher Eigenwert von D_1 mit dem normierten Eigenvektor x_0. Dann läßt sich jeder Vektor $x \in \mathfrak{D}(D^p)$, $p = 1, 2, \dots$, in der Form*

$$x = (x, x_0)\, x_0 + \sum_{k=0}^{p-1} \sum_{j=1}^{n} (U_j D^k x)\, T^k q_j + T^p D^p x \tag{27}$$

darstellen.

Wegen $\|T^p D^p x\| \leqq |\lambda_1|^{-p} \|D^p x\|$ für $x \in \mathfrak{D}_\infty$ und $p = 1, 2, \ldots$ ergibt sich aus (27):

Satz 7: *Jedes Element $x \in \mathfrak{D}_\infty$ mit der Eigenschaft*

$$\lim_{p\to\infty} |\lambda_1|^{-p} \|D^p x\| = 0 \tag{28}$$

läßt sich als Taylor*sche Reihe darstellen:*

$$x = (x, x_0)\, x_0 + \sum_{k=0}^{\infty} \sum_{j=1}^{n} (U_j D^k x)\, T^k q_j \,. \tag{29}$$

Analog wie Satz 2 läßt sich Satz 7 verallgemeinern:

Satz 8: *Jedes Element $x \in \mathfrak{D}_\infty$ mit der Eigenschaft*

$$\lim_{p\to\infty} |\lambda_{m+1}|^{-p} \|D^p x\| = 0 \tag{30}$$

für ein gewisses $m \geqq 1$ läßt sich in eine konvergente Taylor*sche Reihe*

$$x = F_m x + \sum_{k=1}^{\infty} \sum_{j=1}^{n} (U_j D^k x)\, T_m^k q_j \tag{31}$$

entwickeln, wobei $T_m\colon \mathfrak{B} \to \mathfrak{B}$ der durch (11) *erklärte Operator ist und $F_m = I - T_m D\colon \mathfrak{D} \to \mathfrak{B}$ der auf $\mathfrak{D}$ erklärte Operator*

$$F_m x = \sum_{k=0}^{m} (x, x_k)\, x_k + \sum_{j=1}^{n} (U_j x) \left[q_j - \sum_{k=1}^{m} (q_j, x_k)\, x_k \right] \tag{32}$$

ist. Ferner gilt

$$T_m^k q_j = T^k q_j - \sum_{i=1}^{m} \lambda_i^{-k} (q_j, x_i)\, x_i \,. \tag{33}$$

Als Folgerung aus Satz 8 ergibt sich, daß jedes Element $x \in \mathfrak{D}_\infty$ mit der Eigenschaft (30) für gewisses $m \geqq 1$ und $R_j D^k x = 0$, $j = 1, 2, \ldots, n$; $k = 0, 1, \ldots$, eine Linearkombination der Eigenvektoren $x_0, x_1, \ldots, x_m$ ist.

Wir gehen nun zu Anwendungsbeispielen der drei beschriebenen Methoden über und beginnen mit den bekanntesten Interpolationsverfahren.

4. Interpolationsverfahren nach Lagrange

Wir wählen $\mathfrak{B}$ als den Vektorraum $C[a, b]$ der auf $[a, b]$, $-\infty < a < b < \infty$, komplexwertigen stetigen Funktionen und $t_j \in [a, b]$, $j = 1, 2, \ldots, n$, mit $t_j \neq t_k$ für $j \neq k$. Es seien $D = \mathrm{d}^n/\mathrm{d}t^n$ und $R_j x(t) = x(t_j)$, $j = 1, 2, \ldots, n$. Wir wenden die erste Methode an. Dann bilden die Funktionen

$$u_j(t) = \prod_{\substack{k=1\\k\neq j}}^{n} (t - t_k)\,, \qquad j = 1, 2, \ldots, n\,,$$

eine Basis von $\mathfrak{N}$. Die zugehörige (n, n)-Matrix $\boldsymbol{G}$ hat Diagonalform

$$\boldsymbol{G} = \operatorname{diag}\big(u_1(t_1), u_2(t_2), \ldots, u_n(t_n)\big)\,,$$

so daß deren Inverse

$$\boldsymbol{G}^{-1} = \operatorname{diag}\big(u_1(t_1)^{-1}, u_2(t_2)^{-1}, \ldots, u_n(t_n)^{-1}\big)$$

ist. Dann lautet der Randprojektor $F\colon \mathfrak{D} \to \mathfrak{N}$

$$Fx(t) = \sum_{j=1}^{n} \frac{u_j(t)}{u_j(t_j)}\, x(t_j)\,.$$

Eine Rechtsinverse $T_0\colon \mathfrak{B} \to \mathfrak{B}$ von D ist

$$T_0 x(t) = \int_a^t \frac{(t-\tau)^{n-1}}{(n-1)!}\, x(\tau)\, \mathrm{d}\tau\,,$$

so daß der Green*sche* Operator $T\colon \mathfrak{B} \to \mathfrak{B}$ die Form (4) hat. Für $x(t) \in \mathfrak{D}$ gilt nun

$$TDx(t) = \int_a^t \frac{(t-\tau)^{n-1}}{(n-1)!}\, x^{(n)}(\tau)\, \mathrm{d}\tau - \sum_{j=1}^{n} \frac{u_j(t)}{u_j(t_j)} \int_a^{t_j} \frac{(t_j-\tau)^{n-1}}{(n-1)!}\, x^{(n)}(\tau)\, \mathrm{d}\tau = \sum_{j=1}^{n} \frac{u_j(t)}{u_j(t_j)} \int_{t_j}^{t} \frac{(t_j-\tau)^{n-1}}{(n-1)!}\, x^{(n)}(\tau)\, \mathrm{d}\tau$$

oder, wie man durch mehrmalige Anwendung des Satzes von Rolle auf die Hilfsfunktion

$$TDx(t) - \gamma \prod_{k=1}^{n} (t - t_k)\,, \qquad \gamma \in \mathscr{C}\,,$$

zeigt,

$$TDx(t) = \frac{x^{(n)}(\xi)}{n!} \prod_{k=1}^{n} (t - t_k)$$

mit einem (von t abhängigen) Wert $\xi \in (a, b)$.

Für $x(t) \in \mathfrak{D}$ ergibt sich damit das *Interpolationsverfahren von* Lagrange

$$x(t) = Fx(t) + TDx(t)$$

oder

$$x(t) = \sum_{j=1}^{n} \frac{u_j(t)}{u_j(t_j)} x(t_j) + \frac{x^{(n)}(\xi)}{n!} \prod_{k=1}^{n} (t - t_k)$$

mit $\xi \in (a, b)$.

5. Interpolationsverfahren nach Newton

Unter den gleichen Voraussetzungen wie in 4. wählen wir jetzt die Funktionen

$$u_1(t) = 1\,, \qquad u_k(t) = \prod_{j=1}^{k-1} (t - t_j)\,, \qquad k = 2, 3, \dots, n\,,$$

als Basis von $\mathfrak{N}$. Wir wenden erneut die erste Methode an. Die zugehörige (n, n)-Matrix G hat Dreiecksform und lautet

$$G = \begin{pmatrix} 1 & 0 & 0 & & 0 \\ 1 & t_2 - t_1 & 0 & & 0 \\ 1 & t_3 - t_1 & (t_3 - t_1)(t_3 - t_2) & & 0 \\ \vdots & \vdots & \vdots & & \vdots \\ 1 & t_n - t_1 & (t_n - t_1)(t_n - t_2) & \dots & (t_n - t_1) \cdot \ldots \cdot (t_n - t_{n-1}) \end{pmatrix}.$$

Eine einfache Rechnung liefert den Randprojektor

$$Fx(t) = \sum_{j=1}^{n} [t_j, \dots, t_2, t_1]\, u_j(t)$$

mit

$$[t_j] = x(t_j)\,, \qquad j = 1, 2, \dots, n\,,$$

$$[t_{j+1}, t_j, \dots, t_2, t_1] = \frac{[t_{j+1}, t_j, \dots, t_2] - [t_j, \dots, t_2, t_1]}{t_{j+1} - t_1}.$$

Der Greensche Operator T ist dann durch (4) bestimmt.

Für $x(t) \in \mathfrak{D}$ ergibt sich damit das *Interpolationsverfahren von* Newton

$$x(t) = Fx(t) + TDx(t)$$

oder

$$x(t) = \sum_{j=1}^{n} [t_j, \dots, t_2, t_1]\, u_j(t) + \frac{x^{(n)}(\xi)}{n!} \prod_{k=1}^{n} (t - t_k)$$

mit einem (von t abhängigen) Wert $\xi \in (a, b)$.

Weitere Interpolationsverfahren erhält man durch spezielle Wahl der Punkte t_j (siehe z. B. [10], S. 72–83).

6. Interpolationsverfahren nach Hermite

Wir wählen $\mathfrak{B} = C[a, b]$ und $-\infty < a \leqq t_j \leqq b < \infty$, $j = 1, 2, \dots, n$, mit $t_j \neq t_k$ für $j \neq k$. Es seien $r_1, r_2, \dots, r_n$ natürliche Zahlen mit $r_1 + r_2 + \dots + r_n = N$. Dann betrachten wir $D = \mathrm{d}^N/\mathrm{d}t^N$ und die Funktionale $R_{jk}x(t) = x^{(k)}(t_j)$, $j = 1, 2, \dots, n$; $k = 0, 1, \dots, r_j - 1$. Man nennt t_j einen r_j-fachen Interpolationsknoten. Mit Hilfe von

$$p_j(t) = \prod_{\substack{k=1 \\ k \neq j}}^{n} (t - t_k)^{r_k}\,, \qquad j = 1, 2, \dots, n\,,$$

erklärt man die Polynome

$$u_{jk}(t) = (t - t_j)^k\, p_j(t)\,, \qquad j = 1, 2, \dots, n;\ k = 0, 1, \dots, r_j - 1\,,$$

die eine Basis von $\mathfrak{N}$ bilden.

Wir wenden die erste Methode an. Die (N, N)-Matrix

$$G = \begin{pmatrix} R_{10}u_{10} & \dots & R_{10}u_{n, r_n - 1} \\ \vdots & & \vdots \\ R_{n, r_n - 1}u_{10} & \dots & R_{n, r_n - 1}u_{n, r_n - 1} \end{pmatrix}$$

ist eine verallgemeinerte Diagonalmatrix

$$G = \operatorname{diag}(G_1, G_2, \dots, G_n)$$

mit (r_j, r_j)-Dreiecksmatrizen G_j, $j = 1, 2, \dots, n$, der Form

$$G_j = \begin{pmatrix} p_j(t_j) & 0 & \dots & 0 \\ p_j'(t_j) & p_j(t_j) & \dots & 0 \\ \vdots & \vdots & & \vdots \\ p_j^{(r_j-1)}(t_j) & (r_j - 1)\, p_j^{(r_j-2)}(t_j) & \dots & (r_j - 1)!\, p_j(t_j) \end{pmatrix}.$$

Sind $q_j(t) = (p_j(t))^{-1}$, $j = 1, 2, \dots, n$, so gilt für $j = 1, 2, \dots, n$

$$G_j^{-1} = \begin{pmatrix} q_j(t_j) & 0 & \dots & 0 \\ q_j'(t_j) & q_j(t_j) & \dots & 0 \\ \vdots & \vdots & & \vdots \\ \frac{q_j^{(r_j-1)}(t_j)}{(r_j - 1)!} & \frac{q_j^{(r_j-2)}(t_j)}{(r_j - 2)!\,1!} & \dots & \frac{q_j(t_j)}{(r_j - 1)!} \end{pmatrix}$$

und damit

$$G^{-1} = \operatorname{diag}(G_1^{-1}, G_2^{-1}, \dots, G_n^{-1}).$$

Der Randprojektor $F: \mathfrak{D} \to \mathfrak{N}$ lautet dann für $x(t) \in \mathfrak{D}$

$$Fx(t) = \sum_{j=1}^{n} p_j(t) \sum_{k=0}^{r_j-1} x^{(k)}(t_j) \sum_{m=0}^{r_j-1-k} \frac{(t - t_j)^{m+k}}{m!\,k!}\, q_j^{(m)}(t_j).$$

Führt man mittels des Operators $T_0: \mathfrak{B} \to \mathfrak{B}$

$$T_0x(t) = \int_a^t \frac{(t - \tau)^{N-1}}{(N - 1)!}\, x(\tau)\, d\tau,$$

der eine Rechtsinverse von D ist, noch den GREENschen Operator $T: \mathfrak{B} \to \mathfrak{B}$ durch (4) ein, so erhält man das *Interpolationsverfahren mit mehrfachen Knoten nach* HERMITE: Für $x(t) \in \mathfrak{D}$ gilt

$$x(t) = Fx(t) + TDx(t)$$

oder

$$x(t) = \sum_{j=1}^{n} p_j(t) \sum_{k=0}^{r_j-1} x^{(k)}(t_j) \sum_{m=0}^{r_j-1-k} \frac{(t - t_j)^{m+k}}{m!\,k!}\, q_j^{(m)}(t_j) + \frac{x^{(N)}(\xi)}{N!} \prod_{j=1}^{n} (t - t_j)^{r_j}$$

mit einem (von t abhängigen) Wert $\xi \in (a, b)$.

7. Entwicklung nach Bernoullischen Polynomen

Es sei $\mathfrak{B}$ der HILBERT-Raum $L_2(0, 1)$ der auf $[0, 1]$ komplexwertigen, quadratisch absolut integrierbaren Funktionen und $\mathfrak{D}$ die Menge aller auf $[0, 1]$ absolut stetigen Funktionen $x(t) \in \mathfrak{B}$ mit $x'(t) \in \mathfrak{B}$. Ferner sei $D = i\, d/dt: \mathfrak{D} \to \mathfrak{B}$ und $R_1x(t) = x(1) - x(0)$ für $x(t) \in \mathfrak{D}$. Für $\mathfrak{D}_1 = \{x(t) \in \mathfrak{D};\, x(0) = x(1)\}$ ist $D_1 = D \mid \mathfrak{D}_1$ ein selbstadjungierter Operator mit dem reinen Punktspektrum $\{\lambda_k = 2k\pi;\ k = 0, \pm 1, \dots\}$. Sämtliche Eigenwerte λ_k sind einfach, und zugeordnete normierte Eigenfunktionen sind $x_k(t) = \exp(-2k\pi i t)$, $k = 0, \pm 1, \dots$. Da $\lambda_0 = 0$ ein einfacher Eigenwert von D_1 ist, wenden wir die dritte Methode an.

Der Kern von D ist eindimensional. Somit gilt $U_1 = R_1$ und

$$q_1(t) = t - \tfrac{1}{2} = B_1(t)$$

mit dem ersten BERNOULLIschen Polynom $B_1(t)$. Folglich lautet der Operator $F: \mathfrak{D} \to \mathfrak{N} \oplus \operatorname{span} q_1(t)$

$$Fx(t) = \int_0^1 x(\tau)\, d\tau + (x(1) - x(0))\,(t - \tfrac{1}{2})$$

für $x(t) \in \mathfrak{D}$.

Da $T_0: \mathfrak{B} \to \mathfrak{B}$

$$T_0x(t) = (-i) \int_0^t x(\tau)\, d\tau$$

eine Rechtsinverse von D ist, ist nach (4) der Operator $T: \mathfrak{B} \to \mathfrak{B}$

$$Tx(t) = (-i) \left[\int_0^t x(\tau)\, d\tau - \int_0^1 (t - \tau + \tfrac{1}{2})\, x(\tau)\, d\tau \right]$$

eine verallgemeinerte Inverse und eine DRAZIN-Inverse von D_1 mit dem Index 1. Ferner ist der GREENsche Operator T vollstetig und selbstadjungiert und besitzt die Norm $||T|| = (2\pi)^{-1}$. Durch vollständige Induktion erhält man für $x(t) \in \mathfrak{B}$ und $k = 1, 2, \ldots$

$$T^k x(t) = (-i)^k \left[\int_0^t \frac{(t-\tau)^{k-1}}{(k-1)!} x(\tau)\, d\tau + \frac{(-1)^{k+1}}{k!} \int_0^1 B_k(\tau - t)\, x(\tau)\, d\tau \right] =$$

$$= -(-i)^k \frac{1}{k!} \int_0^1 P_k(t-\tau)\, x(\tau)\, d\tau$$

und

$$T^k q_1(t) = (-i)^k \frac{1}{(k+1)!} B_{k+1}(t)\,,$$

wobei $B_k(t)$ das k-te BERNOULLIsche Polynom und $P_k(t)$ die periodische Fortsetzung (mit der Periode 1) des auf $[0, 1]$ eingeschränkten BERNOULLIschen Polynoms $B_k(t)$ ist. Wir erinnern, daß $B_k(t)$ rekursiv erklärt ist durch

$$B_0(t) = 1\,, \qquad B_k'(t) = kB_{k-1}(t)\,, \qquad B_k(1) - B_k(0) = \delta_{1k}\,, \qquad k = 1, 2, \ldots .$$

Dann ergibt sich aus der TAYLORschen Formel (27) die Entwicklung nach BERNOULLIschen Polynomen einer Funktion $x(t) \in \mathfrak{D}(D^p)$, $p = 1, 2, \ldots$,

$$x(t) = \int_0^1 x(\tau)\, d\tau + \sum_{k=0}^{p-1} [x^{(k)}(1) - x^{(k)}(0)] \frac{1}{(k+1)!} B_{k+1}(t) - \frac{1}{p!} \int_0^1 P_p(t-\tau)\, x^{(p)}(\tau)\, d\tau\,. \tag{34}$$

Nach Satz 7 läßt sich jede Funktion $x(t) \in \mathfrak{D}_\infty$ mit

$$\lim_{p\to\infty} (2\pi)^{-p} ||x^{(p)}(t)|| = 0$$

in eine *Reihe nach BERNOULLIschen Polynomen* entwickeln:

$$x(t) = \int_0^1 x(\tau)\, d\tau + \sum_{k=0}^{\infty} \left(x^{(k)}(1) - x^{(k)}(0)\right) \frac{1}{(k+1)!} B_{k+1}(t)\,.$$

Allgemeiner läßt sich nach Satz 8 jede Funktion $x(t) \in \mathfrak{D}_\infty$ mit

$$\lim_{p\to\infty} \left(2\pi(m+1)\right)^{-p} ||x^{(p)}(t)|| = 0 \tag{35}$$

für eine feste natürliche Zahl m in eine Reihe der Form

$$x(t) = \int_0^1 x(\tau)\, d\tau + \sum_{k=0}^{\infty} \left(x^{(k)}(1) - x^{(k)}(0)\right) C_{k+1}(t) + \sum_{j=1}^{m} (\alpha_j \cos 2\pi jt + \beta_j \sin 2\pi jt)$$

entwickeln, wobei $C_{k+1}(t)$ eine Summe eines trigonometrischen Polynoms mit $B_{k+1}(t)/(k+1)!$ und α_j, β_j komplexe Zahlen sind. Die expliziten Ausdrücke für $C_{k+1}(t)$, α_j und β_j kann man den Formeln (32)—(33) entnehmen. Insbesondere besitzt jede Funktion $x(t) \in \mathfrak{D}_\infty$ mit (35) und $x^{(k)}(1) = x^{(k)}(0)$, $k = 0, 1, \ldots$, die Gestalt

$$x(t) = \alpha_0 + \sum_{j=1}^{m} (\alpha_j \cos 2\pi jt + \beta_j \sin 2\pi jt)$$

mit gewissen $\alpha_0, \alpha_j, \beta_j \in \mathscr{C}$, $j = 1, 2, \ldots$.

Die Entwicklungsformel (34) läßt sich leicht verallgemeinern. Dazu seien a, b und m beliebige ganze Zahlen mit $a < b$ und $m \geqq 0$. Ist $y(s)$ auf $[a, b]$ definiert und $(2m+1)$-mal stetig differenzierbar und ist c eine beliebige ganze Zahl mit $a \leqq c < b$, so kann man $y(s)$ auf $[c, c+1]$ einschränken und auf die Funktion $x(t) = y(c+t)$ für $t \in [0, 1]$ die Formel (34) anwenden. Dann erhält man für $s \in [c, c+1]$ die Gleichung

$$y(s) - \int_c^{c+1} y(\sigma)\, d\sigma + \sum_{k=0}^{2m} \left(y^{(k)}(c+1) - y^{(k)}(c)\right) \frac{B_{k+1}(s-c)}{(k+1)!} - \frac{1}{(2m+1)!} \int_c^{c+1} P_{2m+1}(s-\sigma)\, y^{(2m+1)}(\sigma)\, d\sigma\,. \tag{36}$$

Ohne Mühe läßt sich aus (36) die modifizierte EULERsche Summenformel herleiten ([2], S. 105, 109). Es sei $0 \leqq \varepsilon < 1$ vorgegeben. Setzt man $s = c + \varepsilon$ in (36), wobei man für c der Reihe nach die Werte $a, a+1, \ldots, b-1$ wählt, und addiert man anschliessend alle $(b-a)$ Gleichungen, so erhält man die *modifizierte EULERsche Summenformel*

$$\sum_{=a}^{b-1} y(k+\varepsilon) = \int_a^b y(\sigma)\, d\sigma + \sum_{k=0}^{2m} \left(y^{(k)}(b) - y^{(k)}(a)\right) \frac{B_{k+1}(\varepsilon)}{(k+1)!} - \frac{1}{(2m+1)!} \int_a^b P_{2m+1}(\varepsilon - \sigma)\, y^{(2m+1)}(\sigma)\, d\sigma\,.$$

Speziell für $\varepsilon = 1/2$ ergibt sich wegen $B_k(1/2) = 0$ für ungerades k die Formel

$$\sum_{k=a}^{b-1} y\left(k + \frac{1}{2}\right) = \int_a^b y(\sigma)\,d\sigma + \sum_{j=1}^{m} \left(y^{(2j-1)}(b) - y^{(2j-1)}(a)\right) \frac{B_{2j}(1/2)}{(2j)!} - \frac{1}{(2m+1)!} \int_a^b P_{2m+1}\left(\frac{1}{2} - \sigma\right) y^{(2m+1)}(\sigma)\,d\sigma.$$

Im Fall $\varepsilon = 0$ bekommt man die bekannte Eulersche *Summenformel.*

8. Entwicklung nach Eulerschen Polynomen

Wie in 7. seien $\mathfrak{B}$, $\mathfrak{D}$ und D gewählt. Weiterhin sei $R_1x(t) = x(0) + x(1)$ für $x(t) \in \mathfrak{D}$ und

$$\mathfrak{D}_1 = \{x(t) \in \mathfrak{D};\, x(0) + x(1) = 0\}\,.$$

Dann ist $D_1 = D \mid \mathfrak{D}_1$ ein selbstadjungierter Operator mit dem reinen Punktspektrum

$$\{\lambda_k = (2k+1)\,\pi;\, k = 0, \pm 1, \ldots\}\,.$$

Alle Eigenwerte λ_k sind einfach und zugehörige normierte Eigenfunktionen sind $x_k(t) = \exp\left(-(2k+1)\,\pi i t\right)$, $k = 0, \pm 1, \ldots$. Da Null kein Eigenwert von D_1 ist, können wir die zweite Methode anwenden.

Der Kern von D ist eindimensional. Wählt man $u_1(t) = 1$, so lautet wegen $R_1u_1(t) = 2$ der Randprojektor $F\colon \mathfrak{D} \to \mathfrak{N}$

$$Fx(t) = \tfrac{1}{2}\left(x(0) + x(1)\right).$$

Mittels der Rechtsinversen $T_0\colon \mathfrak{B} \to \mathfrak{B}$ von D mit

$$T_0x(t) = (-i) \int_0^t x(\tau)\,d\tau$$

bestimmt man nach (4) den Greenschen Operator $T\colon \mathfrak{B} \to \mathfrak{B}$

$$Tx(t) = (-i)\left[\int_0^t x(\tau)\,d\tau - \tfrac{1}{2}\int_0^1 x(\tau)\,d\tau\right].$$

Der Operator T ist eine Inverse von D_1 und eine Rechtsinverse von D. Ferner ist T selbstadjungiert und vollstetig, und es gilt $\|T\| = \pi^{-1}$. Durch vollständige Induktion erhält man für $x(t) \in \mathfrak{B}$ und $k = 1, 2, \ldots$

$$T^kx(t) = (-i)^k \left[\int_0^t \frac{(t-\tau)^{k-1}}{(k-1)!}\,x(\tau)\,d\tau - \frac{1}{2}\int_0^1 \frac{E_{k-1}(1+t-\tau)}{(k-1)!}\,x(\tau)\,d\tau\right]$$

und

$$T^k\frac{1}{2} = (-i)^k \frac{1}{2k!} E_k(t)\,,$$

wobei $E_k(t)$ das k-te Eulersche Polynom ist, welches rekursiv durch die Gleichungen

$$E_0(t) = 1, \qquad E_k'(t) = kE_{k-1}(t)\,, \qquad E_k(0) + E_k(1) = 0\,, \qquad k = 1, 2, \ldots,$$

erklärt wird.

Damit ergibt sich aus der Taylorschen Formel (6) die *Entwicklung nach* Euler*schen Polynomen* einer Funktion $x(t) \in \mathfrak{D}(D^p)$, $p = 1, 2, \ldots$,

$$x(t) = \frac{1}{2}\sum_{k=0}^{p-1} \left(x^{(k)}(0) + x^{(k)}(1)\right) \frac{1}{k!} E_k(t) + \int_0^t \frac{(t-\tau)^{p-1}}{(p-1)!}\,x^{(p)}(\tau)\,d\tau - \frac{1}{2}\int_0^1 \frac{E_{p-1}(t-\tau+1)}{(p-1)!}\,x^{(p)}(\tau)\,d\tau\,.$$

Nach Satz 2 läßt sich jede Funktion $x(t) \in \mathfrak{D}_\infty$ mit

$$\lim_{p\to\infty} \pi^{-p}\,\|x^{(p)}(t)\| = 0$$

in eine *Reihe nach* Euler*schen Polynomen* entwickeln:

$$x(t) = \frac{1}{2}\sum_{k=0}^{\infty} \left(x^{(k)}(0) + x^{(k)}(1)\right) \frac{1}{k!} E_k(t)\,.$$

9. Entwicklung nach Lidstoneschen Polynomen

Es sei $\mathfrak{B} = L_2(0, 1)$ und $\mathfrak{D}$ die Menge aller auf $[0, 1]$ erklärten Funktionen $x(t) \in \mathfrak{B}$, für die $x(t)$ und $x'(t)$ absolut stetig auf $[0, 1]$ sind und $x''(t) \in \mathfrak{B}$ ist. Ferner sei $D = -d^2/dt^2\colon \mathfrak{D} \to \mathfrak{B}$ und $R_1x(t) = x(0)$, $R_2x(t) = x(1)$ für $x(t) \in \mathfrak{D}$. Für $\mathfrak{D}_1 = \{x(t) \in \mathfrak{D};\, x(0) = x(1) = 0\}$ ist $D_1 = D \mid \mathfrak{D}_1$ ein selbstadjungierter Operator mit dem reinen Punktspektrum $\{\lambda_k = k^2\pi^2;\, k = 1, 2, \ldots\}$. Alle Eigenwerte λ_k sind einfach und zugeordnete normierte Eigenfunktionen sind $x_k(t) = \sqrt{2}\sin k\pi t$, $k = 1, 2, \ldots$. Da Null kein Eigenwert von D_1 ist, können wir die zweite Methode anwenden.

Der Kern von D ist zweidimensional. Als Basis von $\mathfrak{N}$ wählen wir $y_1(t) = 1 - t$ und $y_2(t) = t$. Dann ist G die Einheitsmatrix, so daß der Randprojektor $F\colon \mathfrak{D} \to \mathfrak{N}$ nach (3)

$$Fx(t) = (1-t)\,x(0) + tx(1)$$

für $x(t) \in \mathfrak{D}$ lautet. Mit Hilfe der Rechtsinversen $T_0\colon \mathfrak{B} \to \mathfrak{B}$ von D

$$T_0x(t) = -\int_0^t (t-\tau)\,x(\tau)\,d\tau$$

kann man durch (4) den GREENschen Operator $T\colon \mathfrak{B} \to \mathfrak{B}$

$$Tx(t) = -\int_0^1 G(t,\tau)\,x(\tau)\,d\tau$$

mit der GREENschen Funktion

$$G(t,\tau) = \begin{cases} \tau(t-1) & \text{für} \quad 0 \leqq \tau < t\,, \\ t(\tau-1) & \text{für} \quad t \leqq \tau \leqq 1 \end{cases}$$

bestimmen. Mittels vollständiger Induktion erhält man für $k = 1, 2, \ldots$

$$T^k y_1(t) = (-1)^k A_k(1-t)\,, \qquad T^k y_2(t) = (-1)^k A_k(t)$$

mit dem k-ten LIDSTONEschen Polynom $A_k(t)$, welches rekursiv durch die Gleichungen

$$A_0(t) = t\,, \qquad A_k''(t) = A_{k-1}(t)\,, \qquad A_k(0) = A_k(1) = 0\,, \qquad k = 1, 2, \ldots\,,$$

definiert ist.

Auf Grund der TAYLORschen Formel (6) ergibt sich nun für $x(t) \in \mathfrak{D}(D^p)$, $p = 1, 2, \ldots$, die *Entwicklung nach* LIDSTONE*schen Polynomen*:

$$x(t) = \sum_{k=0}^{p-1} \left(x^{(2k)}(0)\,A_k(1-t) + x^{(2k)}(1)\,A_k(t)\right) + (-1)^p\,T^p x^{(2p)}(t)\,.$$

Nach Satz 2 läßt sich folglich jede Funktion $x(t) \in \mathfrak{D}_\infty$ mit

$$\lim_{p\to\infty} \pi^{-p} \|x^{(p)}(t)\| = 0$$

in eine Reihe nach LIDSTONE*schen* Polynomen entwickeln:

$$x(t) = \sum_{k=0}^{\infty} \left(x^{(2k)}(0)\,A_k(1-t) + x^{(2k)}(1)\,A_k(t)\right).$$

Allgemeiner kann man nach Satz 4 jede Funktion $x(t) \in \mathfrak{D}_\infty$ mit

$$\lim_{p\to\infty} (m+1)^{-p}\,\pi^{-p} \|x^{(p)}(t)\| = 0 \tag{37}$$

für eine feste natürliche Zahl m in eine Reihe der Form

$$x(t) = \sum_{k=0}^{\infty} \left(x^{(2k)}(0)\,B_k(1-t) + x^{(2k)}(1)\,B_k(t)\right) + \sum_{j=1}^{m} \beta_j \sin j\pi t$$

mit

$$B_k(t) = A_k(t) + 2 \sum_{j=1}^{m} (-1)^{j+k}\,(j\pi)^{-2k+1} \sin j\pi t\,,$$

$$\beta_j = 2 \int_0^1 x(\tau) \sin j\pi\tau\,d\tau$$

entwickeln. Dieses Ergebnis wurde zuerst von R. C. BUCK ([4], S. 15) bewiesen. Als Folgerung ergibt sich ein Resultat von I. J. SCHOENBERG ([4], S. 16), wonach jede Funktion $x(t) \in \mathfrak{D}_\infty$ mit (37) und $x^{(2k)}(0) = x^{(2k)}(1) = 0$, $k = 0, 1, \ldots$, gleich einem trigonometrischen Polynom ist.

10. Entwicklung nach Hermiteschen Funktionen

Wir betrachten in $\mathfrak{B} = L_2(-\infty, \infty)$ die sogenannte HERMITEsche oder temperierte Ableitung $D = \frac{d}{dt} + \frac{t}{2}$, die auf der Menge $\mathfrak{D}$ aller absolut stetigen Funktionen $x(t) \in \mathfrak{B}$ mit $Dx(t) \in \mathfrak{B}$ erklärt ist. Der Kern von D ist eindimensional und wird von $\exp(-t^2/4)$ aufgespannt. Es sei $R_1x(t) = \left(x(t), \exp(-t^2/4)\right)$. Wir benutzen die erste Methode, weil Null kein Eigenwert von D_1 ist. Wegen

$$\int_{-\infty}^{\infty} \exp\left(-\frac{t^2}{2}\right) dt = \sqrt{2\pi}$$

ist

$$y_1(t) = \frac{1}{\sqrt{2\pi}} \exp\left(-\frac{t^2}{4}\right).$$

Der Operator $F\colon \mathfrak{B} \to \mathfrak{N}$ projiziert $\mathfrak{B}$ orthogonal auf $\mathfrak{N}$, denn es gilt

$$Fx(t) = \frac{1}{\sqrt{2\pi}} \exp\left(-\frac{t^2}{4}\right) \int_{-\infty}^{\infty} \exp\left(-\frac{\tau^2}{4}\right) x(\tau)\, d\tau$$

für $x(t) \in \mathfrak{B}$. Eine Rechtsinverse von D ist der beschränkte Operator $T_0\colon \mathfrak{B} \to \mathfrak{B}$ mit

$$T_0 x(t) = \exp\left(-\frac{t^2}{4}\right) \int_0^t \exp\left(\frac{\tau^2}{4}\right) x(\tau)\, d\tau$$

für $x(t) \in \mathfrak{B}$, für dessen Norm $\|T_0\| \leqq \sqrt{5}$ gilt. Denn wie in [1], S. 167 gezeigt wurde, gilt die Abschätzung

$$\|T_0 x(t)\|^2 = \int_{-\infty}^{\infty} \exp\left(-\frac{t^2}{2}\right) \left| \int_0^t \exp\left(\frac{\tau^2}{4}\right) x(\tau)\, d\tau \right| dt \leqq 5\, \|x(t)\|^2$$

für alle $x(t) \in \mathfrak{B}$. Somit gilt für den beschränkten GREENschen Operator $T = T_0 - FT_0\colon \mathfrak{B} \to \mathfrak{B}$

$$Tx(t) = \frac{1}{\sqrt{2\pi}} \exp\left(-\frac{t^2}{4}\right) \int_{-\infty}^{\infty} \exp\left(-\frac{\xi^2}{2}\right) \int_{\xi}^{t} \exp\left(\frac{\tau^2}{4}\right) x(\tau)\, d\tau\, d\xi$$

für $x(t) \in \mathfrak{B}$. Mittels vollständiger Induktion beweist man für $k = 0, 1, \ldots$

$$T^k y_1(t) = (\sqrt{2\pi}\, 2^{k/2} k!)^{-1} H_k\left(\frac{t}{\sqrt{2}}\right) \exp\left(-\frac{t^2}{4}\right) = y_{k+1}(t)$$

mit dem k-ten HERMITEschen Polynom $H_k(t)$, welches bekanntlich durch die Gleichung

$$H_k(t) = (-1)^k \exp t^2 \frac{d^k}{dt^k} \exp(-t^2)\,, \qquad k = 0, 1, \ldots,$$

definiert ist.

Für $x(t) \in \mathfrak{D}(D^k)$, $k = 0, 1, \ldots$, gilt nun

$$R_1 D^k x(t) = \left(\left(\frac{d}{dt} + \frac{t}{2}\right)^k x(t)\,,\ \exp\left(-\frac{t^2}{4}\right)\right) = \left(x(t), \left(-\frac{d}{dt} + \frac{t}{2}\right)^k \exp\left(-\frac{t^2}{4}\right)\right) =$$

$$= \left(x(t), 2^{-k/2} H_k\left(\frac{t}{\sqrt{2}}\right) \exp\left(-\frac{t^2}{4}\right)\right) = \sqrt{2\pi}\, k!\, \big(x(t), y_{k+1}(t)\big)\,.$$

Auf Grund der TAYLORschen Formel (6) erhält man für $x(t) \in \mathfrak{D}(D^p)$, $p = 1, 2, \ldots$, die Darstellung

$$x(t) = \sum_{k=0}^{p-1} \big(R_1 D^k x(t)\big)\, y_{k+1}(t) + T^p D^p x(t) =$$

$$= \sqrt{2\pi} \sum_{k=0}^{p-1} k!\, \big(x(t), y_{k+1}(t)\big)\, y_{k+1}(t) + T^p D^p x(t)\,. \tag{38}$$

Da andererseits

$$\big(y_j(t), y_k(t)\big) = \big(\sqrt{2\pi}(j-1)!\big)^{-1} \delta_{jk}\,, \qquad j, k = 1, 2, \ldots,$$

gilt und die Menge aller Funktionen der Form $\exp(-t^2/4)\, p(t)$ mit beliebigen Polynomen $p(t)$ in $\mathfrak{B}$ dicht liegt ([1], S. 140), bilden die Funktionen $\sqrt[4]{2\pi}\sqrt{k!}\, y_{k+1}(t)$, $k = 0, 1, \ldots$, ein vollständiges Orthonormalsystem in $\mathfrak{B}$. Die TAYLORsche Formel (38) liefert deshalb eine endliche Entwicklung nach einem Orthonormalsystem mit Restglied für Funktionen aus $\mathfrak{D}(D^p)$. Aus der Vollständigkeit des betrachteten Orthonormalsystems folgt die *Reihenentwicklung nach HERMITEschen Funktionen* für beliebige $x(t) \in \mathfrak{B}$:

$$x(t) = \sum_{k=0}^{\infty} \sqrt{2\pi}\, k! \big(x(t), y_{k+1}(t)\big)\, y_{k+1}(t)\,.$$

11. Entwicklung nach Hermiteschen Polynomen

Es sei $\mathfrak{B}$ der HILBERT-Raum $L_{2,\omega(t)}(-\infty, \infty)$ mit der Gewichtsfunktion $\omega(t) = \exp(-t^2)$ und dem Skalarprodukt

$$\big(x(t), y(t)\big) = \int_{-\infty}^{\infty} \exp(-\tau^2)\, x(\tau)\, \overline{y(\tau)}\, d\tau\,.$$

Wir betrachten den Operator $D = \mathrm{d}/\mathrm{d}t$, der auf der Menge $\mathfrak{D}$ aller absolut stetigen Funktionen $x(t) \in \mathfrak{B}$ mit $Dx(t) \in \mathfrak{B}$ erklärt ist. Der Kern von D ist eindimensional und wird von $u_1(t) = 1$ erzeugt. Es sei $R_1x(t) = (x(t), 1)$. Da Null kein Eigenwert von D_1 ist, benutzen wir die erste Methode. Wegen

$$\int_{-\infty}^{\infty} \exp(-\tau^2)\,\mathrm{d}\tau = \sqrt{\pi}$$

ist $y_1(t) = \pi^{-1/2}$. Der Operator $F: \mathfrak{B} \to \mathfrak{N}$ projiziert $\mathfrak{B}$ orthogonal auf $\mathfrak{N}$, denn es gilt

$$Fx(t) = \frac{1}{\sqrt{\pi}} \int_{-\infty}^{\infty} \exp(-\tau^2)\, x(\tau)\,\mathrm{d}\tau$$

für $x(t) \in \mathfrak{B}$. Eine Rechtsinverse von D ist der beschränkte Operator $T_0: \mathfrak{B} \to \mathfrak{B}$ mit

$$T_0x(t) = \int_0^t x(\tau)\,\mathrm{d}\tau$$

für $x(t) \in \mathfrak{B}$. Die Beschränktheit von T_0 ergibt sich aus der Ungleichung

$$||T_0x(t)||^2 = \int_{-\infty}^{\infty} \exp(-t^2) \left| \int_0^t x(\tau)\,\mathrm{d}\tau \right|^2 \mathrm{d}t \leqq \tfrac{1}{2} \int_{-\infty}^{\infty} \exp(-t^2)\,|x(t)|^2\,\mathrm{d}t = \tfrac{1}{2}\,||x(t)||^2$$

für beliebige $x(t) \in \mathfrak{B}$. Die obige Abschätzung folgt aus

$$\int_0^\delta \exp(-t^2) \left| \int_0^t x(\tau)\,\mathrm{d}\tau \right|^2 \mathrm{d}t \leqq \int_0^\delta \exp(-t^2) \left(\int_0^t |x(\tau)|\,\mathrm{d}\tau \right)^2 \mathrm{d}t \leqq \int_0^\delta \exp(-t^2) \left(\int_0^t \mathrm{d}\tau \int_0^t |x(\tau)|^2\,\mathrm{d}\tau \right) \mathrm{d}t \leqq$$

$$\leqq \int_0^\delta t \exp(-t^2) \int_0^t |x(\tau)|^2\,\mathrm{d}\tau\,\mathrm{d}t \leqq \tfrac{1}{2} \int_0^\delta \exp(-t^2)\,|x(t)|^2\,\mathrm{d}t\,,$$

wobei die HÖLDERsche Ungleichung und partielle Integration benutzt wurde, für beliebiges $\delta > 0$ und einer analogen Ungleichung für $\delta < 0$. Somit erhält man für den beschränkten Operator $T = T_0 - FT_0: \mathfrak{B} \to \mathfrak{B}$

$$Tx(t) = \frac{1}{\sqrt{\pi}} \int_{-\infty}^{\infty} \exp(-\xi^2) \int_{\xi}^{t} x(\tau)\,\mathrm{d}\tau\,\mathrm{d}\xi$$

für $x(t) \in \mathfrak{B}$. Für $k = 0, 1, \ldots$ gilt nun

$$T^k y_1(t) = (\sqrt{\pi}\, 2^k k!)^{-1} H_k(t) = y_{k+1}(t)$$

mit dem k-ten HERMITEschen Polynom $H_k(t)$.

Für $x(t) \in \mathfrak{D}(D^k)$, $k = 0, 1, \ldots)$, gilt nun

$$R_1D^kx(t) = \int_{-\infty}^{\infty} \exp(-\tau^2)\, x^{(k)}(\tau)\,\mathrm{d}\tau =$$

$$= (-1)^k \int_{-\infty}^{\infty} x(\tau) \frac{\mathrm{d}^k}{\mathrm{d}\tau^k} \exp(-\tau^2)\,\mathrm{d}\tau = (x(t), H_k(t)) = \sqrt{\pi} 2^k k!\, (x(t), y_{k+1}(t))\,.$$

Auf Grund der TAYLORschen Formel (6) erhält man für $x(t) \in \mathfrak{D}(D^p)$, $p = 1, 2, \ldots$, die Darstellung

$$x(t) = \sum_{j=0}^{p-1} (R_1D^jx(t))\, y_{j+1}(t) + T^pD^px(t) = \sum_{j=0}^{p-1} \sqrt{\pi}\, 2^j j!\, (x(t), y_{j+1}(t))\, y_{j+1}(t) + T^pD^px(t)\,.$$

Da andererseits

$$(y_j(t), y_k(t)) = (\sqrt{\pi}\, 2^{j-1}(j-1)!)^{-1}\, \delta_{jk}$$

für $j, k = 1, 2, \ldots$ gilt und die Menge aller Polynome in $\mathfrak{B}$ dicht liegt, bilden die normierten HERMITEschen Polynome

$$\sqrt{2^j j!}\, \sqrt[4]{\pi}\, y_{j+1}(t) = (\sqrt{2^j j!}\, \sqrt[4]{\pi})^{-1} H_j(t)\,, \qquad j = 0, 1, \ldots,$$

ein vollständiges Orthonormalsystem in $\mathfrak{B}$. Aus der Vollständigkeit dieses Orthonormalsystems folgt die *Reihenentwicklung nach HERMITEschen Polynomen* für beliebige $x(t) \in \mathfrak{B}$:

$$x(t) = \sum_{j=0}^{\infty} \sqrt{\pi}\, 2^j j!\, (x(t), y_{j+1}(t))\, y_{j+1}(t)\,.$$

12. Entwicklung nach harmonischen Funktionen

Es sei $\Omega \subset R^n$, $n \geqq 3$, ein beschränktes Gebiet mit hinreichend glattem Rand $\partial\Omega$. Wir betrachten den im distributionentheoretischen Sinn anzuwendenden LAPLACE-Operator $\Delta = D$ mit $\mathfrak{D} = \{u(x) \in C(\overline{\Omega});\, \Delta u(x) \in C(\overline{\Omega})\}$, wobei

$\overline{\Omega} = \Omega \cup \partial\Omega$ ist. Nach dem bekannten WEYLschen Lemma besteht der Kern $\mathfrak{N}$ von D nur aus harmonischen Funktionen. Ist μ_x das harmonische Maß bezüglich Ω und $F\colon C(\overline{\Omega}) \to \mathfrak{N}$ der durch

$$Fu(x) = \begin{cases} \int\limits_{\partial\Omega} u(y)\, \mathrm{d}\mu_x(y)\,, & x \in \Omega\,, \\ u(x)\,, & x \in \partial\Omega\,, \end{cases}$$

erklärte Randprojektor, so hat das innere DIRICHLET-Problem

$$Du(x) = 0\,, \quad x \in \Omega; \quad u(x) = g(x)\,, \quad x \in \partial\Omega$$

bei vorgegebenem $g(x) \in C(\partial\Omega)$ die schwache Lösung $u(x) = Fg(x)$. Mit Hilfe der Fundamentallösung von D

$$E(x) = -\frac{1}{(n-2)\,\omega_n}|x|^{2-n}\,, \quad x \neq o\,,$$

wobei ω_n den Inhalt der n-dimensionalen Einheitskugeloberfläche bezeichnet, und des Randprojektors F kann man die zum inneren DIRICHLET-Problem für D bezüglich Ω gehörende Rechtsinverse $T\colon C(\overline{\Omega}) \to C(\overline{\Omega})$ bestimmen:

$$Tu(x) = \int\limits_{\Omega} E(x-y)\, u(y)\, \mathrm{d}y - \Big(F\int\limits_{\Omega} E(x-y)\, u(y)\, \mathrm{d}y\Big) = \int\limits_{\Omega} \Gamma(x, y)\, u(y)\, \mathrm{d}y\,,$$

wobei

$$\Gamma(x, y) = \begin{cases} -\dfrac{1}{(n-2)\,\omega_n}\left(|x-y|^{2-n} - \displaystyle\int\limits_{\partial\Omega} |z-y|^{2-n}\, \mathrm{d}\mu_x(z)\right), & x \in \Omega\,, \quad y \in R^n\,, \quad x \neq y\,, \\ 0, & x \in \partial\Omega\,, \quad y \in R^n\,, \end{cases}$$

die entsprechende GREENsche Funktion ist. Es gilt dann

$$DTu(x) = u(x)\,, \quad x \in \Omega; \quad Tu(x) = 0\,, \quad x \in \partial\Omega\,,$$

für alle $u(x) \in C(\overline{\Omega})$ und

$$u(x) = Fu(x) + TDu(x)\,, \quad x \in \Omega\,,$$

für alle $u(x) \in \mathfrak{D}$. Die Beweise und weitere Einzelheiten findet man z. B. in [13], S. 42–71.

Es sei nun für $k = 1, 2, \ldots$

$$\Gamma_1(x, y) = \Gamma(x, y)\,, \quad \Gamma_{k+1}(x, y) = \int\limits_{\Omega} \Gamma_k(x, z)\, \Gamma(z, y)\, \mathrm{d}z\,,$$

so daß man für die Potenzen $T^k\colon C(\overline{\Omega}) \to C(\overline{\Omega})$

$$T^k u(x) = \int\limits_{\Omega} \Gamma_k(x, y)\, u(y)\, \mathrm{d}y$$

erhält.

Wir wenden nun auf eine beliebige Funktion $u(x) \in \overline{C}^{2p}(\Omega)$, $p = 1, 2, \ldots$, d. h., $u(x)$ ist mit allen ihren partiellen Ableitungen bis einschließlich $2p$-ter Ordnung stetig in Ω und stetig fortsetzbar auf $\overline{\Omega}$, die TAYLORsche Formel (1) an und erhalten die Darstellung ([11])

$$u(x) = \sum_{j=0}^{p-1} T^j h_j(x) + T^p D^p u(x)\,, \quad x \in \Omega\,, \tag{39}$$

mit eindeutig bestimmten harmonischen Funktionen $h_j(x) = FD^j u(x)$, $j = 0, 1, \ldots, p-1$. Gilt für $u(x) \in \overline{C}^{2p}(\Omega)$ außerdem $D^p u(x) \equiv 0$ für $x \in \Omega$, so bekommt man die folgende *Darstellung einer p-polyharmonischen Funktion* ([11])

$$u(x) = \sum_{j=0}^{p-1} T^j h_j(x)\,, \quad x \in \Omega\,.$$

Ist $u(x) \in \overline{C}^{\infty}(\Omega)$, d. h., $u(x)$ ist mit allen ihren partiellen Ableitungen stetig in Ω und stetig fortsetzbar auf $\overline{\Omega}$, und gilt für alle $x \in \Omega$ und jedes $j = 0, 1, \ldots$

$$|D^j u(x)| \leqq M < \infty\,,$$

so läßt sich $u(x)$ in eine absolut und gleichmäßig konvergente Reihe der Form

$$u(x) = \sum_{j=0}^{\infty} T^j h_j(x)\,, \quad x \in \Omega\,, \tag{40}$$

mit eindeutig bestimmten harmonischen Funktionen $h_j(x)$, $j = 0, 1, \ldots$, entwickeln, falls Ω in einer Kugel $K(x_0, r)$ mit dem Mittelpunkt $x_0 \in R^n$ und einem Radius $r < \sqrt{2n}$ enthalten ist ([11], [12]).

Denn für

$$w(x) = \int\limits_{\Omega} \Gamma(x, y)\, \mathrm{d}y$$

gilt $Dw(x) = 1$, $x \in \Omega$, und $w(x) = 0$, $x \in \partial\Omega$, so daß

$$v(x) = w(x) - \frac{|x - x_0|^2}{2n}$$

harmonisch ist. Aus dem Maximum-Prinzip harmonischer Funktionen und der Voraussetzung $\Omega \subseteq K(x_0, r)$ folgt dann

$$|w(x)| = \left| \int_\Omega \Gamma(x, y)\, \mathrm{d}y \right| \leqq \frac{r^2}{2n}, \qquad x \in \overline{\Omega}.$$

Durch vollständige Induktion zeigt man allgemeiner für $k = 1, 2, \ldots$

$$\left| \int_\Omega \Gamma_k(x, y)\, \mathrm{d}y \right| \leqq \left(\frac{r^2}{2n}\right)^k, \qquad x \in \overline{\Omega}.$$

Nach dem Maximum-Prinzip harmonischer Funktionen gilt aufgrund der Voraussetzung für $j = 0, 1, \ldots$

$$|FD^j u(x)| \leqq M,$$

und somit ergibt sich für alle $x \in \Omega$ und $j = 1, 2, \ldots$

$$|T^j FD^j u(x)| = \left| \int_\Omega \Gamma_j(x, y)\, FD^j u(y)\, \mathrm{d}y \right| \leqq M \left| \int_\Omega \Gamma_j(x, y)\, \mathrm{d}y \right| \leqq M \left(\frac{r^2}{2n}\right)^j.$$

Analog erhält man

$$|T^j D^j u(x)| \leqq M \left(\frac{r^2}{2n}\right)^j.$$

Wegen $r < \sqrt{2n}$ ist die Reihe

$$\sum_{j=1}^{\infty} M \left(\frac{r^2}{2n}\right)^j$$

eine konvergente Majorante der Reihe in Formel (40) mit $h_j(x) = FD^j u(x) \in \mathfrak{N}$, die wegen (39) die Summe $u(x)$ besitzt.

Literatur

1 ANTOSIK, P.; MIKUSIŃSKI, J.; SIKORSKI, R., Theory of Distributions, The Sequential Approach, Elsevier Scientific Publ. Comp. Amsterdam — PWN Warszawa 1973.

2 BERG, L., Operatorenrechnung. II: Funktionentheoretische Methoden, DVW Berlin 1974.

3 BERG, L., Randwertprobleme in der Operatorenrechnung, Wiss. Z. Univ. Rostock **23**: M 8 (1974), 621—624.

4 BOAS, R. P.; BUCK, R. C., Polynomial Expansions of Analytic Functions, Springer-Verlag Berlin-Göttingen-Heidelberg 1958.

5 BUCKHOLTZ, J. D.; SHAW, J. K., Generalized completely convex functions and STURM-LIOUVILLE operators, SIAM J. Math. Analysis **6**: 5 (1974), 812—828.

6 NASHED, M. Z.; VOTRUBA, G. V., A unified operator theory of generalized inverses. In: NASHED, M. Z. (Ed.), Generalized Inverses and Applications, Proc. Advanced Sem. "Generalized Inverses and Applications", Wisconsin-Madison 8—10 Oct. 1973, Academic Press New York-San Francisco-London 1976, S. 1—109.

7 SHAW, J. K., Series expansions and linear differential operators, SIAM J. Math. Analysis **7**: 3 (1976), 311—331.

8 TASCHE, M., Randprojektoren und GREENsche Operatoren von abstrakten Randwertproblemen und deren Anwendung, ZAMM **57** (1977), 579—589.

9 TASCHE, M., Zur Konvergenzbeschleunigung von FOURIER-Reihen, Math. Nachr. **90** (1979), 113—134

10 WILLERS, F. A., Methoden der praktischen Analysis, Walter de Gruyter & Co. Berlin 1950.

11 NICOLESCU, M., Unele proprietăţi de descompunere ale funcţiilor de mai multe variabile reale şi în particular ale funcţiilor poliarmonice, Revista Univ. C. I. Parhon 4—5 (1954), 53—63.

12 NICOLESCU, M., Problème de l'analyticité par rapport à un opérateur linéaire, Studia Math. **16** (1958), 353—363.

13 SCHULZE, B.-W.; WILDENHAIN, G., Methoden der Potentialtheorie für elliptische Differentialgleichungen beliebiger Ordnung, Akademie-Verlag, Berlin 1977.

Eingereicht: 29. 3. 1978

Anschrift: Doz. Dr. sc. nat. MANFRED TASCHE, Wilhelm-Pieck-Universität Rostock, Sektion Mathematik, Universitätsplatz 1, DDR-2500 Rostock

KLEINE MITTEILUNGEN

ZAMM 61, 395 – 396 (1981)

F. BEICHELT

A Replacement Policy Based on Limits for the Repair Cost Rate

1. Introduction

A common replacement policy for technical systems consists in applying the economic lifetime, i.e. the system is replaced by a new one at that moment in which the expected loss cost rate (*expected total loss cost per unit time*) is minimum.

The strict application of the economic lifetime does not allow taking into account individual deviations of repair costs of specific systems from the average cost development. Thus the *repair limit replacement policy* has been suggested [5]. This policy is characterized as follows: When a failure happens, the system is first inspected and the necessary repair cost is estimated. If this cost exceeds a certain amount, called "repair limit", then the system is not repaired but replaced. Examples show that the application of optimum repair limits via using the economic lifetime leads to significant cost savings [3].

A *disadvantage* of the repair limit replacement policy is the decision "replace or not" depending only on the cost of a single repair. Long-lasting situations characterized by frequent happenings of repairs whose costs are below the corresponding limits have, however, no direct influence on the moment of replacement, although the total repair cost per unit time (*repair cost rate*) might justify a replacement. Thus, further financial savings seem to be possible if the replacement-decision depends on the whole history of the maintenance process. This leads to the following replacement policy:

The system is replaced as soon as the repair cost rate $Z(t)$ reaches or exceeds a critical level Z_0.

To the knowledge of the author this policy has not been treated in the literature so far, never mind its practicability and its simple structure. (There exists a formal relationship to policies considered e.g. in [1, 6], where replacements are scheduled according to the realizations of a so-called "prognostic parameter".)

Basic Assumptions

1. Infinite planning horizon.
2. Replacements and repairs take only negligible times (stated only for numerical convenience).

2. Mathematical analysis

Let $A(t)$ be the random total repair cost during time interval $(0, t)$ and let the repair cost rate $Z(t) = A(t)/t$ be a strictly increasing, continuous function of t almost everywhere, $t > 0$. (In realty, because of assumption 2 the realizations of $A(t)$ are expected to be piecewise linear functions of t with jumps at the moments of repair.) According to the policy the system is replaced at the moment Y given by $Z(Y) = Z_0$. Let Y_i be the random time between the two subsequent replacements $i - 1$ and i, $i = 1, 2, \ldots$. By definition of the policy, $Z(Y_i) = Z_0$. The random variables Y_i are assumed to be independent and identically distributed as Y with $\mathsf{E}(Y) < \infty$.

The expected loss cost rate is (loss cost = repair + replacement cost)

$$K(Z_0) = \frac{C}{\mathsf{E}(Y)} + Z_0, \tag{1}$$

where C is the cost of a replacement.

Proof: Within the interval $\left(0, \sum_{i=1}^{n} Y_i\right)$ the loss cost rate amounts to

$$\frac{nC + \left(\sum_{i=1}^{n} Y_i\right) Z_0}{\sum_{i=1}^{n} Y_i} = \frac{C}{\frac{1}{n}\sum_{i=1}^{n} Y_i} + Z_0 .$$

Letting $n \to \infty$ and applying the strong law of the large numbers yields equation (1).

For simplicity, in what follows let us assume $A(t) = Bt + Xt^2$, where X is a positive random variable with $\mathsf{E}(1/X) < \infty$ and B is a constant with $0 \leq B < Z_0$. Hence the repair cost rate is

$$Z(t) = B + Xt . \tag{2}$$

This equation implies a linear expected repair cost rate $M(t) = \mathsf{E}(Z(t)) = B + \mathsf{E}(X)\, t$ which has been proved being true in many practical situations [1].

A *repair limit which is optimum with respect to* (1) is denoted by Z_0^*, i.e. it is

$$K(Z_0^*) = \min_{Z_0 \in (0,\infty)} K(Z_0) .$$

In view of (2) there holds $Y = (Z_0 - B)/X$. Therefore,

$$K(Z_0) = \frac{C}{(Z_0 - B)\,\mathsf{E}(1/X)} + Z_0 ,$$

$$Z_0^* = \sqrt{\frac{C}{\mathsf{E}(1/X)}} + B , \qquad K(Z_0^*) = 2\sqrt{\frac{C}{\mathsf{E}(1/X)}} + B .$$

Let $Q(T)$ be the expected loss cost rate, if the replacement interval T is used:

$$Q(T) = \frac{\mathsf{E}(A(T)) + C}{T} = \frac{C}{T} + M(T) .$$

The *economic lifetime* T^* is defined by

$$Q(T^*) = \min_{T \in (0,\infty)} Q(T) .$$

Simple calculations yield

$$T^* = \sqrt{\frac{C}{\mathsf{E}(X)}} , \qquad Q(T^*) = 2\sqrt{C\mathsf{E}(X)} + B .$$

In view of $1/\mathsf{E}(X) < \mathsf{E}(1/X)$, there holds the following theorem being a justification of the replacement policy treated.

Theorem: *Applying the optimum repair limit Z_0^* is more effective than applying the economic lifetime, i.e. it is*

$$K(Z_0^*) < Q(T^*) .$$

(Assuming the existence of Z_0^* and T^* the theorem is expected to be true for all $M(t)$ which are increasing and convex functions of t.)

3. Special case

Let be X uniformely distributed in the interval (a, b), $0 < a < b < \infty$, i.e. it is

$$\mathsf{P}(X < x) = \begin{cases} 0, & 0 < x \leq a \\ \dfrac{x - a}{b - a}, & a < x \leq b \\ 1, & x > b . \end{cases}$$

Then,

$$\mathsf{P}(Y > t) = \begin{cases} 1, & 0 < t = \dfrac{Z_0 - B}{b} \\ \dfrac{Z_0 - B - at}{t(b - a)}, & \dfrac{Z_0 - B}{b} < t \leq \dfrac{Z_0 - B}{a} \\ 0, & \dfrac{Z_0 - B}{a} < t , \end{cases}$$

$$\mathsf{E}(Y) = \int_0^\infty \mathsf{P}(Y > t)\, \mathrm{d}t = \frac{Z_0 - B}{b - a} \ln \frac{b}{a},$$

$$K(Z_0) = \frac{(b - a)\, C}{(Z_0 - B) \ln \dfrac{b}{a}} + Z_0 ,$$

$$Z_0^* = \sqrt{\frac{(b - a)\, C}{\ln \dfrac{b}{a}}} + B , \qquad K(Z_0^*) = 2Z_0^* - B .$$

The corresponding characteristic quantities for the economic lifetime-method are

$$Q(T) = \frac{C}{T} + \frac{a+b}{2} T + B\,,$$

$$T^* = \sqrt{\frac{2C}{a+b}}\,,$$

$$Q(T^*) = 2\sqrt{\frac{(a+b)\,C}{2}} + B\,.$$

Numerical example: Let be $C = 6900$, $B = 526$ and $\mathsf{E}(X) = \frac{a+b}{2} = 100$. Figure 1 shows the per cent gain

$$\Delta K = \frac{Q(T^*) - K(Z_0^*)}{Q(T^*)}\,100\%$$

depending on $u = b - a$, $0 \leq u < 2\mathsf{E}(X) = 200$. If u is sufficiently large this gain exceeds 25%!

For the same average repair cost development and replacement cost as in this example the application of optimum repair limits for each repair instead of using the economic lifetime leads to cost savings coming to 6% [2, 3].

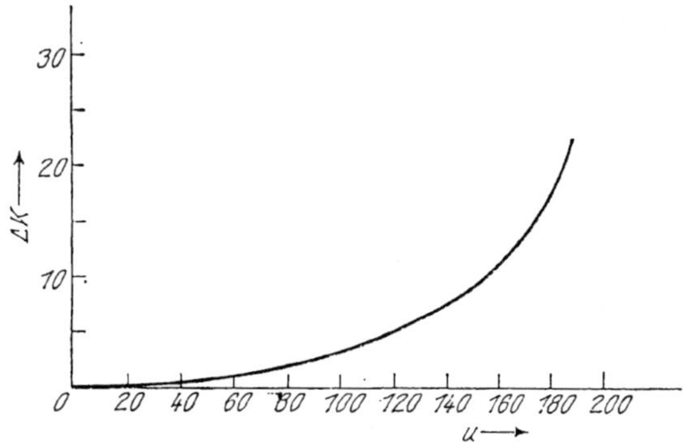

Fig. 1. Comparision of effectivity of repair limit replacement policy and economic lifetime method

References

1 Beichelt, F., Effektive Planung prophylaktischer Maßnahmen in der Instandhaltung, Berlin, VEB Verlag Technik 1979.
2 Beichelt, F.; Fischer, K., General failure model applied to preventive maintenance policies, IEEE Trans. on Reliability R-29, No. 1, 39–41 (1980).
3 Drinkwater, R. W.; Hastings, N. V., An economic replacement model, Operat. Res. Quarterly 18, No. 2, 121–138 (1967).
4 Družinin, G. W., Reliability of Automatic Systems, Moscow, Energija, 3th ed. 1977, in Russian.
5 Gardent, P.; Nonat, L., Entretien et renouvellement d'un parc de machines, Revue franç. Rech. operat., No. 7, 5–19 (1963).
6 Gertsbakh, I. B., Models of Preventive Maintenance, Amsterdam-New York, North-Holland-Publishing Company 1977.

Eingereicht: 12. 6. 1980, revidierte Fassung: 22. 12. 1980

Anschrift: Doz. Dr. F. Beichelt, Hochschule für Verkehrswesen „Friedrich List", Friedrich-List-Platz 1, DDR-8010 Dresden

ZAMM 61, 396–399 (1981)

L. Berg

Stabile Iterationsverfahren beliebiger Ordnung zur Berechnung von Wurzeln

Von den Diskretisierungsverfahren für gewöhnliche Differentialgleichungen ist bekannt, daß eine Steigerung der Genauigkeit in der Regel nur auf Kosten der Stabilität möglich ist. Ganz entsprechend erfordern Überrelaxationsverfahren zur Lösung von Gleichungssystemen bei positiven Iterationsmatrizen den Übergang zu negativen Elementen in der Hauptdiagonale, so daß bei Rechnungen auf einem Automaten in kritischen Fällen Stellenauslöschungen auftreten und die theoretische Genauigkeitssteigerung nicht realisiert werden kann.

Aus diesem Grunde ist es bemerkenswert, daß es zur Berechnung n-ter Wurzeln Iterationsverfahren

$$x_{m+1} = g(x_m) \tag{1}$$

beliebiger Ordnung gibt, die nur positive Koeffizienten besitzen und damit positive Wurzeln numerisch stabil zu berechnen gestatten.

Im folgenden wird eine Klasse solcher Iterationsverfahren angegeben und auf einige ihrer Eigenschaften untersucht. Da die guten Konvergenzeigenschaften nur in einer hinreichend kleinen Umgebung der gesuchten Wurzel zum Tragen kommen, sind die Verfahren bei praktischer Anwendung mit anderen Verfahren zu koppeln, die geeignete Startwerte liefern. Auf andersartige Iterationsverfahren (vgl. [5]), speziell auf Mehrstufenverfahren (vgl. [2]), sowie auf direkte Berechnungsmethoden für Wurzeln (vgl. [4]) gehen wir hier nicht weiter ein. Statt dessen wird zum Abschluß noch der Übergang zum Fall beliebiger Polynomwurzeln angedeutet.

Satz: *Das Iterationsverfahren* (1) *zur Berechnung von* $\sqrt[n]{a}$ *mit* $a \neq 0$, $n \geq 2$, *besitzt im Fall*

$$g(x) = \frac{\sum\limits_{j=0}^{k} \binom{k-\frac{1}{n}}{j}\binom{k+\frac{1}{n}}{k-j} x^{nj+1} a^{-j}}{\sum\limits_{j=0}^{k} \binom{k-\frac{1}{n}}{k-j}\binom{k+\frac{1}{n}}{j} x^{nj} a^{-j}} \tag{2}$$

die Konvergenzordnung $2k+1$ *und den Konvergenzfaktor*

$$c = (-1)^{k-1}\binom{k-\frac{1}{n}}{2k+1}\frac{k!^2}{(2k)!}\, n^{2k+1} a^{-2k/n}\,, \tag{3}$$

sowie im Fall

$$g(x) = \frac{\sum\limits_{j=0}^{k} \binom{k-\frac{1}{n}}{j}\binom{k-1+\frac{1}{n}}{k-j} x^{nj} a^{-j}}{\sum\limits_{j=1}^{k} \binom{k-\frac{1}{n}}{k-j}\binom{k-1+\frac{1}{n}}{j-1} x^{nj-1} a^{-j}} \tag{4}$$

die Konvergenzordnung $2k$ *und den Konvergenzfaktor*

$$c = 2(-1)^k\binom{k-\frac{1}{n}}{2k}\frac{k!^2}{(2k)!}\, n^{2k} a^{(1-2k)/n}\,, \tag{5}$$

wobei k *eine beliebige natürliche Zahl ist und die Aussagen auf jeden der* n *Werte von* $\sqrt[n]{a} = a^{1/n}$ *zutreffen.*

Beweis: Ohne Beschränkung der Allgemeinheit können wir $a = 1$ annehmen, da wir sonst x nur durch $x\sqrt[n]{a}$ zu ersetzen brauchen. Unsere Behauptung geht damit in

$$g(x) - 1 \sim c(x-1)^p$$

für $x \to 1$ mit $p = 2k+1$ im ersten und $p = 2k$ im zweiten Fall über. Im weiteren Verlauf des Beweises verstehen wir unter $x^{1/n}$ stets denjenigen Zweig, der für $x \to 1$ gegen 1 strebt. Dann gilt $\sqrt[n]{x} - 1 \sim \frac{1}{n}(x-1)$ für $x \to 1$, und unsere Behauptung kann durch

$$g(\sqrt[n]{x}) - 1 \sim cn^{-p}(x-1)^p$$

ersetzt werden. Setzen wir weiterhin $g(\sqrt[n]{x}) = Z(x)/N(x)$, so bleibt schließlich

$$Z(x) = N(x)\,(1 + cn^{-p}(x-1)^p + o(x-1)^p)$$

zu zeigen, d. h.

$$Z^{(l)}(1) = N^{(l)}(1) \tag{6}$$

für $l = 0, 1, \ldots, p-1$ und

$$Z^{(p)}(1) = N^{(p)}(1) + cn^{-p}p!\,N(1)\,. \tag{7}$$

Im ersten Fall wählen wir

$$Z(x) = \sum_{j=0}^{k}\binom{k-\frac{1}{n}}{j}\binom{k+\frac{1}{n}}{k-j} x^j\,,$$

$$N(x) = \sum_{j=0}^{k} \binom{k-\frac{1}{n}}{k-j}\binom{k+\frac{1}{n}}{j} x^{j-1/n}\,.$$

Auf Grund der Produktformel für Potenzreihen und der CAUCHYschen Koeffizientenformel für den wir die Integraldarstellungen

$$Z(x) = \frac{1}{2\pi i}\int (1+xz)^{k-1/n}\,(1+z)^{k+1/n}\,z^{-k-1}\,dz\,,$$

$$N(x) = \frac{x^{-1/n}}{2\pi i}\int (1+z)^{k-1/n}\,(1+xz)^{k+1/n}\,z^{-k-1}\,dz\,,$$

wobei jeweils über einen hinreichend kleinen Kreis in der z-Ebene um den Nullpunkt zu integrieren ist. Ersetzen wir z durch z/x, so ergibt sich für das zweite Integral die Darstellung

$$N(x) = \frac{1}{2\pi i}\int (x+z)^{k-1/n}\,(1+z)^{k+1/n}\,z^{-k-1}\,dz\,.$$

Durch Differentiation folgt jetzt

$$Z^{(l)}(1) = \binom{k-\frac{1}{n}}{l}\frac{l!}{2\pi i}\int (1+z)^{2k-l}\,z^{-k+l-1}\,dz$$

$$= \binom{k-\frac{1}{n}}{l}\,l!\binom{2k-l}{k-l},$$

$$N^{(l)}(1) = \binom{k-\frac{1}{n}}{l}\frac{l!}{2\pi i}\int (1+z)^{2k-l}\,z^{-k-1}\,dz$$

$$= \binom{k-\frac{1}{n}}{l}\,l!\binom{2k-l}{k},$$

so daß aus $\binom{2k-l}{k-l} = \binom{2k-l}{k}$ die Behauptung (6) und aus (3) die Behauptung (7) mit $p = 2k + 1$ hervorgeht.

Im zweiten Fall wählen wir

$$Z(x) = \sum_{j=0}^{k}\binom{k-\frac{1}{n}}{j}\binom{k-1+\frac{1}{n}}{k-j} x^{j},$$

$$N(x) = \sum_{j=1}^{k}\binom{k-\frac{1}{n}}{k-j}\binom{k-1+\frac{1}{n}}{j-1} x^{j-1/n}$$

und erhalten wegen

$$z(1+z)^{q} = \sum_{j=1}^{\infty}\binom{q}{j-1} z^{j}$$

mit $q = k - 1 + 1/n$ zunächst die Integraldarstellungen

$$Z(x) = \frac{1}{2\pi i}\int (1+xz)^{k-1/n}\,(1+z)^{k-1+1/n}\,z^{-k-1}\,dz\,,$$

$$N(x) = \frac{x^{1-1/n}}{2\pi i}\int (1+z)^{k-1/n}\,z(1+xz)^{k-1+1/n}\,z^{-k-1}\,dz$$

und nach einer entsprechenden Umformung wie zuvor

$$N(x) = \frac{1}{2\pi i}\int (x+z)^{k-1/n}\,(1+z)^{k-1+1/n}\,z^{-k}\,dz\,.$$

Hieraus folgen die Gleichungen

$$Z^{(l)}(1) = \binom{k-\frac{1}{n}}{l}\frac{l!}{2\pi i}\int (1+z)^{2k-l-1}\,z^{-k+l-1}\,dz =$$

$$= \binom{k-\frac{1}{n}}{l}\,l!\binom{2k-l-1}{k-l},$$

$$N^{(l)}(1) = \binom{k-\frac{1}{n}}{l}\frac{l!}{2\pi i}\int (1+z)^{2k-l-1}\,z^{-k}\,dz =$$

$$= \binom{k-\frac{1}{n}}{l}\,l!\binom{2k-l-1}{k-1}$$

und damit auch die Behauptungen (6) und (7) mit (5) und $p = 2k$.

Es sei betont, daß nicht nur die Binomialkoeffizienten in (2) und (4), sondern auch die *Konvergenzfaktoren* (3) *und* (5) *für positive Wurzeln stets positiv* sind.

Spezialfälle: Schreiben wir statt $g(x)$ genauer $g_p(x)$, wobei p die Ordnung des zugehörigen Iterationsverfahrens (1) bezeichnet, so erhalten wir aus (4) bzw. (2) für $k = 1$

$$g_2(x) = \frac{(n-1)\,x^n + a}{n x^{n-1}}\,, \tag{8}$$

$$g_3(x) = \frac{(n-1)\,x^{n+1} + (n+1)\,a x}{(n+1)\,x^n + (n-1)\,a} \tag{9}$$

und für $k = 2$

$$g_4(x) = \frac{(2n-1)\,(n-1)\,x^{2n} + 2(2n-1)\,(n+1)\,a x^n + (n+1)\,a^2}{2n(n+1)\,x^{2n-1} + 2n(2n-1)\,a x^{n-1}}\,, \tag{10}$$

$$g_5(x) = \frac{(2n-1)\,(n-1)\,x^{2n+1} + 2(2n-1)\,(2n+1)\,a x^{n+1} + (2n+1)\,(n+1)\,a^2 x}{(2n+1)\,(n+1)\,x^{2n} + 2(2n-1)\,(2n+1)\,a x^n + (2n-1)\,(n-1)\,a^2}\,. \tag{11}$$

Das zugehörige Iterationsverfahren (1) ist im Fall $g = g_2(x)$ das gewöhnliche NEWTON-Verfahren für die Gleichung $x^n - a = 0$ und im Fall $g = g_3(x)$ mit $n = 2$ ein auf DEDEKIND zurückgehendes Verfahren (vgl. J. TODD [6], S. 37).

Aus den binomischen Formeln

$$x^n = \sum_{j=0}^{n}\binom{n}{j}(x-w)^j\,w^{n-j}\,,$$

$$n x^{n-1} = \sum_{j=1}^{n} j\binom{n}{j}(x-w)^{j-1}\,w^{n-j}$$

ergeben sich leicht die Beziehungen

$$(n-1)\,x^n + w^n - w n x^{n-1} = \sum_{j=2}^{n}(j-1)\binom{n}{j}(x-w)^j\,w^{n-j}\,, \tag{12}$$

$$(n-1)\,x^{n+1} + (n+1)\,w^n x - w((n+1)\,x^n + (n-1)\,w^n) = $$
$$= \sum_{j=3}^{n+1}(j-2)\binom{n+1}{j}(x-w)^j\,w^{n+1-j}\,, \tag{13}$$

aus denen man für $w = \sqrt[n]{a}$ unmittelbar die zugehörigen Konvergenzfaktoren $c = c_p$ entnehmen kann, und zwar

$$c_2 = \frac{n-1}{2w}\,,\qquad c_3 = \frac{n^2-1}{12w^2}\,,$$

die mit (5) bzw. (3) für $k = 1$ übereinstimmen.

Die zu (12) und (13) analogen Beziehungen im Fall $p = 4$

$$Z_4(x) - wN_4(x) = \sum_{j=4}^{2n}((n+1)j + 1 - 5n)\binom{2n}{j}(x-w)^j\,w^{2n-j}\,, \tag{14}$$

wobei $Z_4(x)$ den Zähler und $N_4(x)$ den Nenner von (10) mit $a = w^n$ bezeichnet, sind allerdings nur für $n = 2$ und $n = 3$ richtig, während schon im Fall $n = 4$ auf der rechten Seite von (14) noch der Summand $\binom{8}{4}(x-w)^4\,w^4$ hinzugefügt werden muß. Für die zugehörigen Konvergenzfaktoren $c_4^{(n)}$ können wir wieder ablesen:

$$c_4^{(2)} = \frac{1}{8w^3}\,,\qquad c_4^{(3)} = \frac{5}{9w^3}\,,\qquad c_4^{(4)} = \frac{35}{24w^3} \tag{15}$$

in Übereinstimmung mit den aus (5) für $k = 2$ sich ergebenden Werten

$$c_4^{(n)} = \frac{(2n-1)(n^2-1)}{72\sqrt[n]{a^3}}\,.$$

Die Gleichungen (12) bis (14) eignen sich zur Herleitung von *Fehlerabschätzungen.*

Quadratwurzeln: Im Fall $n = 2$ lassen sich die Funktionen (2) wegen

$$\binom{k-\frac{1}{2}}{j}\binom{k+\frac{1}{2}}{k-j} = \binom{2k+1}{2j+1}\binom{2k}{k}4^{-k}$$

in der Form

$$g_{2k+1}(x) = \frac{\sum\limits_{j=0}^{k}\binom{2k+1}{2j+1}x^{2j+1}a^{k-j}}{\sum\limits_{j=0}^{k}\binom{2k+1}{2j}x^{2j}a^{k-j}} \tag{16}$$

und die Funktionen (4) wegen

$$\binom{k-\frac{1}{2}}{j}\binom{k-\frac{1}{2}}{k-j} = \binom{2k}{2j}\binom{2k}{k}4^{-k},$$

$$\binom{k-\frac{1}{2}}{k-j}\binom{k-\frac{1}{2}}{j-1} = \binom{2k}{2j-1}\binom{2k}{k}4^{-k}$$

in der Form

$$g_{2k}(x) = \frac{\sum\limits_{j=0}^{k}\binom{2k}{2j}x^{2j}a^{k-j}}{\sum\limits_{j=1}^{k}\binom{2k}{2j-1}x^{2j-1}a^{k-j}} \tag{17}$$

schreiben. Bezeichnen wir den Zähler von $g_p(x)$ mit $Z_p(x)$ und den Nenner mit $N_p(x)$, so gehen die beiden vorhergehenden Darstellungen unmittelbar aus

$$Z_p(x) - wN_p(x) = (x-w)^p \tag{18}$$

mit $w = \sqrt{a}$ hervor, und aus (18) lassen sich wegen

$$\sum_{j=0}^{k}\binom{2k+1}{2j} = 2^{2k}, \qquad \sum_{j=1}^{k}\binom{2k}{2j-1} = 2^{2k-1}$$

auch wieder die Konvergenzfaktoren

$$c_{2k+1} = \frac{1}{2^{2k}a^k}, \qquad c_{2k} = \frac{1}{2^{2k-1}a^{k-1/2}}$$

ablesen, die mit (3) bzw. (5) für $n = 2$ übereinstimmen.

Für die Funktionen (16) *und* (17) *gilt*

$$g_2(g_p(x)) = g_{2p}(x) \tag{19}$$

und für $q \geqq 1$

$$g_{2^q}(x) = g_2^{[q]}(x)\,, \tag{20}$$

wobei $g^{[q]}$ *die q-te Iterierte von* $g(x)$ *bezeichnet.*

Zum Beweis von (19) brauchen wir $g_p(x)$ nur in der Form

$$g_p(x) = \frac{(x+w)^p + (x-w)^p}{(x+w)^p - (x-w)^p}\cdot w$$

mit $w = \sqrt{a}$ zu schreiben und an Stelle von x in

$$g_2(x) = \frac{x}{2} + \frac{a}{2x} \tag{21}$$

einzusetzen. Die Gleichung (20) ergibt sich danach aus (19) durch vollständige Induktion.

Die Gleichung (20) besagt, daß die *einmalige Anwendung des Iterationsverfahrens* (1) *der Ordnung* 2^q *mit der Funktion* $g_{2^q}(x)$ *bei der Berechnung von* $\sqrt{a}$ *mit der q-maligen Anwendung des quadratisch konvergenten* NEWTON-*Verfahrens mit der Iterationsfunktion* (21) *identisch ist.* Nebenbei ergibt sich hieraus eine erneute Bestätigung der Konvergenzordnung (vgl. L. COLLATZ [3], S. 231).

Polynomwurzeln: Die vorhergehenden Formeln lassen sich folgendermaßen auf beliebige Polynome übertragen. Will man die Wurzeln w des Polynoms n-ten Grades

$$w^n + a_1 w^{n-1} + \dots + a_n = 0 \tag{22}$$

mit Hilfe eines Iterationsverfahrens (1) berechnen, das für einfache Wurzeln von p-ter Ordnung sein soll, so *gehe man von der Hilfsfunktion*

$$h(x) = \sum_{j=p}^{p+n-2} c_j(x-w)^j \tag{23}$$

aus, ersetze alle Potenzen w^l *mit* $l \geqq n$ *mit Hilfe von* (22) *durch entsprechende Potenzen mit* $l < n$ *und bestimme die* $n-1$ *Unbekannten* c_j *so, daß die neuen Koeffizienten der* w^l *mit* $2 \leqq l \leqq n-1$ *verschwinden,* was in nichttrivialer Weise möglich ist. *Setzt man diese* c_j, die im allgemeinen von x abhängen, aber stets als teilerfremde Polynome gewählt werden können, *in* (23) *ein, so läßt sich* $h(x)$ *in der Form*

$$h(x) = Z(x) - wN(x)$$

mit von w *unabhängigen Funktionen* $Z(x)$ *und* $N(x)$ *schreiben, und die gesuchte Iterationsfunktion* $g(x)$ *lautet*

$$g(x) = \frac{Z(x)}{N(x)}\,. \tag{24}$$

Wie die Gleichungen (12) und (13) zeigen, entstehen bei Anwendung auf das Polynom $w^n - a = 0$ in den Fällen $p = 2$ und $p = 3$ wieder die alten Verfahren. Aber schon im Fall $p = 4$, $n = 3$ findet man

$$c_4 = 10x^3 - a\,, \qquad c_5 = -6x^2$$

und nach weiteren Rechnungen für die Iterationsfunktion (24)

$$g(x) = \frac{4x^7 + 19ax^4 + 4a^2x}{10x^6 + 16ax^3 + a^2}\,, \tag{25}$$

die von der Funktion (10) mit $n = 3$ verschieden ist. Der zugehörige Konvergenzfaktor lautet $\frac{1}{3a}$ und ist daher betragsmäßig kleiner als der Konvergenzfaktor $c_4^{(3)} = \frac{5}{9a}$ aus (15) von $g_4(x)$. Die Funktion (25) muß aber ihrer Bauart nach mit der Funktion

$$g_5(x) = \frac{5x^7 + 35ax^4 + 14a^2x}{14x^6 + 35ax^3 + 5a^2}$$

aus (11) verglichen werden, bei der das Iterationsverfahren (1) von fünfter Ordnung und damit dem entsprechenden Verfahren mit (25) überlegen ist.

Verfahren höherer Ordnung zur Lösung einer Gleichung $f(x) = 0$ kann man bekanntlich aus der TAYLOR-Entwicklung an einer Näherungsstelle x

$$0 = f(x) + \frac{1}{1!}f'(x)\,\Delta x + \frac{1}{2!}f''(x)\,(\Delta x)^2 + \dots \tag{26}$$

gewinnen, indem man beispielsweise auf die Umformung

$$\Delta x = -\frac{f(x)}{f'(x)} - \frac{f''(x)}{2!f'(x)}(\Delta x)^2 - \frac{f'''(x)}{3!f'(x)}(\Delta x)^3 - \dots$$

die Methode der asymptotischen Iteration (vgl. [1]) anwendet. Nach dem NEWTON-Verfahren entsteht dann als nächstes ein Verfahren dritter Ordnung mit der Iterationsfunktion

$$g(x) = x - \frac{f(x)}{f'(x)} - \frac{f''(x)f^2(x)}{2f'^3(x)} \tag{27}$$

usw. (vgl. L. COLLATZ [3], S. 234).

Bildet man von (26) jedoch die Umformung

$$\Delta x = \frac{-f(x)}{f'(x) + \frac{1}{2!}f''(x)\,\Delta x + \frac{1}{3!}f'''(x)\,(\Delta x)^2 + \dots}\,,$$

so entsteht analog nach dem NEWTON-Verfahren ein Verfahren dritter Ordnung mit der Iterationsfunktion

$$g(x) = x - \frac{f(x)f'(x)}{f'^2(x) - \frac{1}{2}f(x)f''(x)}\,. \tag{28}$$

Offenbar ist die Funktion (28) bei Polynomen n-ten Grades einfacher gebaut als (27), und bei dem Beispiel $f(x) = x^n - a$ rech net man leicht nach, daß (28) dann nichts anderes als die bereits bekannte Funktion (9) ist.

Zusatz bei der Korrektur: Nach einem Referat von W. J. KOTZE in den Mathematical Reviews **26 (1963) 3170** über eine Arbeit von B. JANKÓ, Mathematica (Cluj) **2** (25) 269–275 (1960) geht die Formel (28) auf J. S. FRAME, Amer. Math. Monthly **51**, 36–38 (1944) zurück.

Literatur

1 BERG, L., Asymptotische Darstellungen und Entwicklungen, VEB DVW, Berlin 1968.
2 BERG, L., Ein ableitungsfreies Dreistufenverfahren mit quadratischer Konvergenz zur Berechnung von Nullstellen, Rostocker Math. Kolloquium **13**, 119–122 (1980).
3 COLLATZ, L., Funktionalanalysis und Numerische Mathematik, Springer-Verlag, Berlin-Göttingen-Heidelberg 1964.
4 FRANK, E., Continued fraction expansions for k-th roots, ZAMM **60**, T 289–T 291 (1980).
5 SCHWETLICK, H., Numerische Lösung nichtlinearer Gleichungen, VEB DVW, Berlin 1979.
6 TODD, J., Basic Numerical Mathematics **1**: Numerical Analysis, Birkhäuser Verlag, Basel-Stuttgart 1979.

Eingereicht am 6. 1. 1981

Anschrift: Prof. Dr. LOTHAR BERG, Sektion Mathematik der Wilhelm-Pieck-Universität, Universitätsplatz 1, DDR-2500 Rostock

ZAMM 61, 399–401 (1981)

P. H. M. WOLKENFELT

On the Numerical Stability of Reducible Quadrature Methods for Second Kind Volterra Integral Equations

An important class of implicit direct quadrature methods is applied to a two-parameter family of convolution test equations, which includes as a special case the classical test equation. The present analysis provides new insight into the behaviour of numerical methods for VOLTERRA equations, and yields the result that the stability regions of the methods we consider are smaller than the stability region associated with the continuous test model. As an illustration, we give the stability regions of the backward differentiation quadrature methods.

1. Introduction

Consider the VOLTERRA integral equation of the second kind

$$f(x) = g(x) + \int_0^x K(x, y, f(y))\,dy\,, \qquad 0 \leqq x \leqq X\,, \tag{1.1}$$

where $f(x)$ is the unknown function, and where $g(x)$ and the kernel $K(x, y, f)$ are given functions. We assume that the conditions for the existence of a unique continuous solution are satisfied (see e.g. [1, p. 80]).

Direct quadrature methods for the numerical solution of (1.1) are obtained by applying quadrature rules of the form

$$\int_0^{x_n} \varphi(y)\,dy \simeq h \sum_{j=0}^{n} w_{nj}\varphi(x_j)\,, \qquad x_j = jh\,,$$

to discretize (1.1), and yield equations of the form

$$f_n = g(x_n) + h \sum_{j=0}^{n} w_{nj} K(x_n, x_j, f_j)\,, \qquad n \geqq k \geqq 1\,, \tag{1.2}$$

for values f_n approximating $f(x_n)$. Here the value of k depends on the desired accuracy. If the required starting values f_0 $(= g(0)), f_1, \dots, f_{k-1}$ are known, the values $f_k, f_{k+1}, \dots$ can be computed in a step-by-step fashion. BAKER [1, ch. 6] discusses a wide variety of numerical methods for (1.1) including the methods (1.2) for various choices of the weights w_{nj}.

In this paper we consider the class of quadrature methods which are reducible to linear multistep methods for solving ODEs (Ordinary Differential Equations). The construction and analysis of such quadrature methods is treated in [10]. We shall use subsequently an important property of the weights w_{nj}: it holds, for $n \geqq 2k$, that

$$\sum_{i=0}^{k} a_i w_{n-i,j} = \begin{cases} 0 & \text{if } \ 0 \leqq j \leqq n-k-1\,, \\ b_{n-j} & \text{if } \ n-k \leqq j \leqq n\,, \end{cases} \tag{1.3}$$

where we have defined $w_{nj} = 0$ for $j > n$. The real constants a_i and b_i $(i = 0(1)\,k)$ are the coefficients of a convergent linear multistep method.

We assume that $a_0 \neq 0$. From the theory of linear multistep methods for ODEs (see e.g. [8, p. 30]) we recall the characteristic polynomials ϱ and σ:

$$\varrho(\zeta) := \sum_{i=0}^{k} a_i \zeta^{k-i}\,, \qquad \sigma(\zeta) := \sum_{i=0}^{k} b_i \zeta^{k-i}\,, \tag{1.4}$$

the consistency conditions $\varrho(1) = 0$ and $\varrho'(1) = \sigma(1)$, and the fact that ϱ and σ have no common factors. Furthermore, the polynomial ϱ is assumed to satisfy the root condition, that is $\varrho(\zeta) = 0$ implies $|\zeta| \leqq 1$ and $\varrho(\zeta) = \varrho'(\zeta) = 0$ implies $|\zeta| < 1$. The quadrature method (1.2) associated with the linear multistep method (ϱ, σ) through (1.3) is said to be *(ϱ, σ)-reducible.*

The stability behaviour (for fixed $h \neq 0$ and $n \to \infty$) of various methods of the form (1.2) has been analyzed by BAKER and KEECH [2] with respect to the test equation

$$f(x) = 1 + \lambda \int_0^x f(y)\,dy\,. \tag{1.5}$$

It can be derived (see [10]) that the stability behaviour with respect to (1.5) of (ϱ, σ)-reducible quadrature methods is determined by the roots of the characteristic equation

$$\varrho(\zeta) - h\lambda\sigma(\zeta) = 0\,. \tag{1.6}$$

The stability behaviour is well-known from the ODE-theory, since (1.6) is identical to the characteristic equation of the linear multistep method (ϱ, σ) applied to the ODE test equation $f' = \lambda f$ (to which (1.5) is equivalent). Thus the stability analysis based on (1.5) is straightforward, which is a consequence of the fact that the kernel in (1.5) is independent of x.

The main purpose of this paper is to analyze the stability behaviour of (ϱ, σ)-reducible quadrature methods with respect to the convolution test equation

$$f(x) = 1 + \int_0^x \{\lambda + \mu(x-y)\}\,f(y)\,dy\,, \qquad \lambda, \mu \in \mathbb{R}\,, \tag{1.7}$$

which is an extension of (1.5). The choice of this test equation was motivated by the requirement that the kernel in (1.7) depends upon x on the one hand. On the other hand we choose the convolution, because VOLTERRA integral equations of the convolution type occur frequently in applications such as demography [7] and renewal theory [6]. The stability analysis based on (1.7) yields important new insight which cannot be obtained from (1.5). As an example we prove (in § 3) that (ϱ, σ)-reducible quadrature methods which are A-stable with respect to (1.5) are not even A_0-stable with respect to (1.7).

We emphasize that the analyses based on (1.5) and (1.7) are only preliminary steps towards the eventual aim, which is a stability analysis for *discrete* VOLTERRA *equations*

$$f_n = g_n + h \sum_{j=0}^{n} w_{nj} K(x_n, x_j) f_j\,, \qquad n \geqq 0\,. \tag{1.2'}$$

However, at present the theory of discrete VOLTERRA equations is largely non-existent, and will need substantial development before such a general stability analysis can be attempted.

This paper is a shortened version of the institute report [11] and the reader is referred to the report for further detail.

2. Stability analysis with respect to the convolution test equation

By differentiating (1.7) twice, it is readily seen that the solution of the convolution test equation is identical to the solution of the second order differential equation

$$f'' = \lambda f' + \mu f\,, \qquad f(0) = 1\,, \qquad f'(0) = 0\,. \tag{2.1}$$

Application of the direct quadrature method (1.2) to the test equation (1.7) yields the equations

$$f_n = 1 + h\lambda \sum_{j=0}^{n} w_{nj} f_j + h^2\mu \sum_{j=0}^{n} w_{nj}(n-j) f_j . \tag{2.2}$$

Instead of differentiating twice, as we did in the continuous case, we now apply twice a differencing technique to the equations (2.2). We take a weighted sum of successive equations (2.2) to obtain

$$\sum_{i=0}^{k} a_i f_{n-i} = h\lambda \sum_{i=0}^{k} b_i f_{n-i} + h^2\mu \sum_{j=0}^{n} \sum_{i=0}^{k} a_i w_{n-i,j}(n-i-j) f_j =$$

$$= h\lambda \sum_{i=0}^{k} b_i f_{n-i} + h^2\mu \sum_{i=0}^{k} i b_i f_{n-i} - h^2\mu \sum_{j=0}^{n} \sum_{i=0}^{k} i a_i w_{n-i,j} f_j , \tag{2.3}$$

where we have used (1.3) and the equality $\varrho(1) = \sum a_i = 0$ (recall that, by definition, $w_{nj} = 0$ for $j > n$). Notice that for $\mu = 0$, we obtain the difference equation associated with the test equation (1.5).

Applying the same differencing technique again to successive equations (2.3) we find

$$\sum_{l=0}^{k} a_l \sum_{i=0}^{k} a_i f_{n-i-l} =$$

$$= h\lambda \sum_{l=0}^{k} a_l \sum_{i=0}^{k} b_i f_{n-i-l} + h^2\mu \sum_{l=0}^{k} a_l \sum_{i=0}^{k} i b_i f_{n-i-l} -$$

$$- h^2\mu \sum_{j=0}^{n} \sum_{i=0}^{k} \sum_{l=0}^{k} i a_i a_l w_{n-i-l,j} f_j .$$

Using (1.3) again, we finally obtain after some manipulations the following recurrence relation for the numerical solution f_n

$$\sum_{l=0}^{k} a_l \left\{ \sum_{i=0}^{k} a_i - b_i[h\lambda + h^2\mu(i-l)] \right\} f_{n-i-l} = 0 . \tag{2.4}$$

We observe that the solution $f(x)$ of (1.7) (or equivalently (2.1)) tends to zero as $x \to \infty$ if and only if both λ and μ are negative. In particular, if $\lambda^2 + 4\mu > 0$ ($\lambda, \mu < 0$) the exact solution is exponentially decaying whereas for $\lambda^2 + 4\mu < 0$ ($\lambda, \mu < 0$) the solution is a damped oscillation. Ideally, a numerical method for (1.7) ought to preserve this qualitative behaviour of the exact solution. However, since such a requirement may be too strong for a particular method we content ourselves with the requirement that the numerical solution has the same asymptotic behaviour (for $x \to \infty$) as the continuous solution. These observations suggest the following definitions.

Definition 1: The quadrature method (1.2) applied to (1.7) is said to be *absolutely stable* for a given $h\lambda$ and $h^2\mu$ if, for these values of $h\lambda$ and $h^2\mu$, the solution f_n of (2.4) tends to zero as $n \to \infty$.

Definition 2: A region $\mathcal{R}$ in the $(h\lambda, h^2\mu)$-plane is said to be the *region of absolute stability* of the method (1.2) if (1.2) is absolutely stable for all $(h\lambda, h^2\mu) \in \mathcal{R}$.

Definition 3: The quadrature method (1.2) applied to (1.7) is said to be A_0*-stable* if its region of absolute stability includes the third quadrant, that is, if

$$\{(h\lambda, h^2\mu) \mid h\lambda < 0, h^2\mu < 0\} \subset \mathcal{R} .$$

We remark that the definitions given above bear a resemblance to the stability definitions associated with methods for Volterra integro-differential equations (cf. [3, 9]). In Remark 3.2 we shall discuss this connection in more detail.

It is easily verified that the characteristic equation associated with the difference equation (2.4) reads

$$\varrho(\zeta)\,[\varrho(\zeta) - h\lambda\sigma(\zeta)] - h^2\mu \sum_{l=0}^{k} a_l \zeta^{k-l} \sum_{i=0}^{k} b_i (i-l)\, \zeta^{k-i} = 0 . \tag{2.5}$$

We shall refer to the left-hand side of (2.5) as the *stability polynomial* of the method, and denote it by $S(\zeta; h\lambda, h^2\mu)$. This polynomial can be expressed completely in terms of the polynomials ϱ and σ and their first derivatives; to be specific

$$S(\zeta; h\lambda, h^2\mu) := \varrho^2(\zeta) - h\lambda\varrho(\zeta)\,\sigma(\zeta) - h^2\mu\zeta[\sigma(\zeta)\,\varrho'(\zeta) - \varrho(\zeta)\,\sigma'(\zeta)] . \tag{2.6}$$

From definition 1 the reader may verify that the method (1.2) is absolutely stable for a given $h\lambda$ and $h^2\mu$ if and only if the roots ζ_i of $S(\zeta; h\lambda, h^2\mu) = 0$ satisfy $|\zeta_i| < 1$ for $i = 1(1)\,2k$. Therefore, explicit conditions for absolute stability can be derived from (2.6), for example by means of the Routh-Hurwitz or the Schur criterion [8, § 3.7]. As an illustration we derive, in the following examples, the stability regions of two simple quadrature methods.

Example 1 (the repeated trapezium rule): For the quadrature method (1.2) we choose the repeated trapezium rule ($w_{n0} = w_{nn} = 1/2$, $w_{nj} = 1$ for $j = 1(1)\,n-1$), which is reducible to the trapezoidal rule for solving ODEs. For this rule, the polynomials ϱ and σ are $\varrho(\zeta) = \zeta - 1$ and $\sigma(\zeta) = (\zeta + 1)/2$. Application of the Routh-Hurwitz criterion to (2.6) yields the stability region

$$\mathcal{R} = \{(h\lambda, h^2\mu) \mid h\lambda < 0, -4 < h^2\mu < 0\} .$$

Note that the trapezoidal rule, in view of definition 3, is *not* A_0-stable.

Example 2 (the repeated rectangle rule): We now choose the repeated rectangle rule ($w_{n0} = 0$, $w_{nj} = 1$ for $j = 1(1)\,n$) which is reducible to the first order backward differentiation (BD) method or backward Euler rule for solving ODEs ($\varrho(\zeta) = \zeta - 1$ and $\sigma(\zeta) = \zeta$). Application of the Routh-Hurwitz criterion yields the stability region

$$\mathcal{R} = \{(h\lambda, h^2\mu) \mid h\lambda < 0, 2h\lambda - 4 < h^2\mu < 0\} \cup$$

$$\cup \{(h\lambda, h^2\mu) \mid h\lambda > 2, 0 < h^2\mu < 2h\lambda - 4\} .$$

Again we observe that this quadrature method is *not* A_0-stable.

The above examples illustrate a general result which is given in the following section.

3. A result concerning the existence of A_0-stable quadrature methods

One should realize that the notion of A_0-stability loses much in significance if no A_0-stable methods can be found at all. However, we emphasize that A_0-stable methods do exist: Recently, Brunner et al. [4] succeeded in deriving a class of A_0-stable Runge-Kutta methods of Bel'tyukov type. For the class of direct quadrature methods considered in this paper, however, we have the following result.

Theorem 1: *(ϱ, σ)-reducible quadrature methods cannot be A_0-stable.*

Proof: Choose, in (2.6), $h = 1$ and a fixed negative λ and define

$$\varrho_1(\zeta) := \varrho(\zeta)\,[\varrho(\zeta) - \lambda\sigma(\zeta)] , \qquad \sigma_1(\zeta) := \zeta[\sigma(\zeta)\,\varrho'(\zeta) - \varrho(\zeta)\,\sigma'(\zeta)] . \tag{3.1}$$

With this definition of ϱ_1 and σ_1, the stability polynomial (2.6) is identical to the stability polynomial of the linear multistep method (ϱ_1, σ_1) applied with stepsize $h = 1$ to the ODE $f' = \mu f$.

The coefficient of ζ^{2k} in $\varrho_1(\zeta)$ is $a_0(a_0 - \lambda b_0)$ and does not vanish for a suitable negative λ (recall that $a_0 \neq 0$). The coefficient of ζ^{2k} in $\sigma_1(\zeta)$ is $b_0 k a_0 - a_0 k b_0 = 0$, and therefore the linear multistep method (ϱ_1, σ_1) is *explicit*. If ϱ_1 and σ_1 have common factors, they are divided out to yield polynomials ϱ_2 and σ_2 with no common factors. The method (ϱ_2, σ_2), however, *remains explicit.* Cryer [5, theorem 3.1] has shown that explicit linear multistep methods for ODEs cannot be A_0-stable. This result implies the existence of a negative μ such that $\varrho_2(\zeta) - \mu\sigma_2(\zeta)$ has zeros outside the open unit disk. For this μ and the suitable choice of λ, the stability polynomial (2.6) has, therefore, zeros outside the open unit disk. Hence we have shown the existence of a point in the third quadrant of the $(h\lambda, h^2\mu)$-plane, which is outside the stability region. Q.E.D.

We end this section with two remarks.

Remarks:

3.1. The result of the above Theorem can also be derived in a more heuristic way. Consider the equation (2.2) for $h\lambda \to 0$ and $h^2\mu$ fixed. In this case, all (ϱ, σ)-reducible quadrature methods for finding f_n become explicit, and, clearly the region of absolute stability cannot have the entire negative $h^2\mu$-axis as part of its boundary.

3.2. We also want to consider the result of Theorem 1 in connection with similar methods for solving Volterra integro-differential equations (VIDE). Consider the class of numerical methods $(\varrho, \sigma\colon Q)$ (see [3, 9, 10]) where (ϱ, σ) is a linear multistep method for ODEs and where Q represents the set of quadrature formulae. Choosing for Q (ϱ, σ)-reducible quadrature formulae, we obtain the class of methods $(\varrho, \sigma; \varrho, \sigma)$.

The stability analysis of such methods is based on the test equation (cf. [3])

$$f'(x) = \lambda f(x) + \mu \int_0^x f(y)\,dy\,. \tag{3.2}$$

Although this test equation is equivalent to the convolution test equation (1.7), *the stability behaviour with respect to* (3.2) *and* (1.7) *of the same basic method* (ϱ, σ) *is different*. To be specific, it is known ([3, 9]) that A_0-stable methods of the form $(\varrho, \sigma; \varrho, \sigma)$ exist for VIDEs. For example, choosing for (ϱ, σ) the trapezoidal rule or the first or second order BD methods (which are A-stable for ODEs) yields an A_0-stable method $(\varrho, \sigma; \varrho, \sigma)$ for VIDEs. However, for the same underlying method (ϱ, σ) used to solve VOLTERRA integral equation of the second kind, *the property of A_0-stability is lost.*

4. The stability regions of the backward differentiation methods

In section 2 we have seen that the stability regions can be determined by applying the ROUTH-HURWITZ criterion to the stability polynomial S defined in (2.6). Since for a linear k-step method the degree of S is $2k$, in general such a procedure becomes increasingly complicated, and, therefore, it is more convenient to determine the stability regions by other means. In [11] we have described a boundary locus method, and as an illustration we give here the stability regions of the quadrature methods which are reducible to the well-known backward differentiation methods for $k = 2, 3, 4$ and 5. (For $k = 1$ the region was given already in example 2.) The coefficients a_i and b_0 ($b_i = 0$ for $i = 1(1)\,k$) are listed in [8, p. 242]. For a discussion of the corresponding quadrature method we refer to [10].

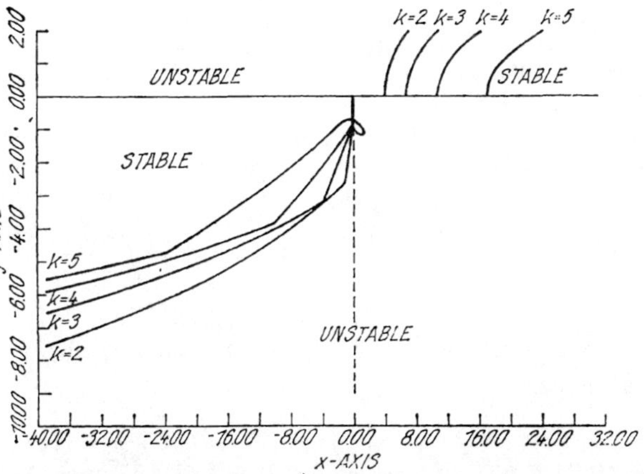

Fig. 1

In the figure above diagrams of the stability regions are given in the (x, y)-plane which represents the $(h\lambda, h^2\mu)$-plane on a different scale. To be specific, $x = h\lambda$ and $y = G(h^2\mu)$ with G defined by $G(z) =$ (if $z \geq 0$ then $\sqrt{z}$ else $-\sqrt{-z}$). Not ethat the scales of the x- and y-axes are different.

For a more detailed discussion on the construction of these regions, the reader is referred to [11], where also a close-up of Figure 1 near the origin can be found.

5. Concluding remarks

In this paper we have investigated the stability behaviour of a class of direct quadrature methods, viz. the (ϱ, σ)-reducible quadrature methods, for the solution of VOLTERRA integral equations of the second kind. An important result is that within this class no A_0-stable methods exist. The question whether there exist more general direct quadrature methods which are A_0-stable, remains open. On the basis of the heuristic argument mentioned in remark 3.1, however, *we conjecture that a step-by-step direct quadrature method cannot be A_0-stable.*

Numerical methods for solving (1.1) which are A_0-stable do exist. Examples are the RUNGE-KUTTA methods discussed in [4]. These methods, however, are outside the class of direct quadrature methods, and therefore do not contradict our conjecture.

References

1 BAKER, C. T. H., The Numerical Treatment of Integral Equations, Oxford, Clarendon Press 1977.
2 BAKER, C. T. H.; KEECH, M. S., Stability regions in the numerical treatment of VOLTERRA integral equations, SIAM J. Numer. Analysis 15 (1978), 394–417.
3 BRUNNER, H.; LAMBERT, J. D., Stability of numerical methods for VOLTERRA integro-differential equations, Computing 12 (1974), 75–89.
4 BRUNNER, H.; NØRSETT, S. P.; WOLKENFELT, P. H. M., On V_0-stability of numerical methods for VOLTERRA integral equations of the second kind, Report NW 84/80, Mathematisch Centrum, Amsterdam 1980.
5 CRYER, C. W., A new class of highly-stable methods: A_0-stable methods, BIT 13 (1973), 153–159.
6 FELLER, W., On the integral equation of renewal theory, Ann. Math. Statist. 12 (1941), 243–267.
7 KEYFITZ, N., Introduction to the Mathematics of Population, Reading, Addison-Wesley 1968.
8 LAMBERT, J. D., Computational Methods in Ordinary Differential Equations, London, John Wiley 1973.
9 MATTHYS, J., A-stable linear multistep methods for VOLTERRA integrodifferential equations, Numer. Math. 27 (1976), 85–94.
10 WOLKENFELT, P. H. M., Linear multistep methods and the construction of quadrature formulae for VOLTERRA integral and integro-differential equations, Report NW 76/79, Mathematisch Centrum, Amsterdam 1979.
11 WOLKENFELT, P. H. M., Stability analysis of reducible quadrature methods for VOLTERRA integral equations of the second kind, Report NW 79/80, Mathematisch Centrum, Amsterdam 1980.

Eingereicht am 2. 7. 1980, revidierte Fassung am 14. 11. 1980

Anschrift: P. H. M. WOLKENFELT, Department of Numerical Mathematics, Stichting Mathematisch Centrum, Kruislaan 413, 1098 SJ Amsterdam, The Netherlands

ZAMM 61, 401–404 (1981)

P. RAJAGOPALAN / R. PURUSHOTHAMAN

Mean Flow Induced by a Travelling Wave in a Roating Fluid

1. Introduction

The response of two dimensional boundary layers to oscillating free streams has been studied extensively by various authors. Excellent reviews of fluctuating flow in boundary layers are given by STEWARTSON [1], STUART [2], RILEY [3], MCCROSKEY [4] and TELIONIS [5]. The effect of fluctuating external flow and that of oscillating boundaries in a rotating fluid was considered by THORNLEY [6]. The fluctuating velocity field due to temporal oscillations of the boundary and fluid has been obtained and the existence of boundary layers of thickness of order $\left(\frac{\nu}{2\Omega \pm \omega}\right)$ has been shown by THORNLEY [6]. The resonance phenomenon that occurs when the forcing frequency ω is equal to the natural frequency of rotation, 2Ω, was analysed by considering an initial value problem and its steady state. The object of the present paper is to study the flow induced in a rotating fluid by an external flow with a velocity in the form of a travelling wave. The quasi-steady flow induced by a propagating wave at large distances from a semi-infinite plate has been discussed by PATEL [7]. The travelling wave introduces oscillating streamwise pressure gradients. When the outer flow in a rotating fluid has only temporal oscillations, the normal velocity is constant and the tangential pressure gradient oscillates with the same frequency of the inviscid flow. On the other hand, we find that the spatially and temporally oscillating disturbances in the far stream introduces tangential pressure gradients with higher harmonics and the normal velocity has spatial and time dependence. As the free stream velocity oscillates with zero mean, the steady normal velocity is zero. The analysis is carried out for the case of large rotation. The velocity functions are taken as a power series in Ro, the ROSSBY number, and the zeroth and first order approximations are obtained. A steady tangential velocity of first order in Ro is obtained. As in the case of THORNLEY [6], no resonance behaviour is found for the critical forcing frequencies.

2. Formulation of the Problem

Consider the motion of a semi-infinite expanse of homogeneous rotating fluid bounded by an infinite plane $z = 0$. The plate and the fluid rotate with a constant angular velocity Ω about the z-axis and an oscillating free stream parallel to the bounding plane is superposed on the steadily rotating flow. The direction of the free stream is taken as the x-axis and the velocity of the external flow is chosen as $U \cos(kx + \omega t)$ where U, k, ω are the amplitude of the disturbance, the wave number and the frequency of oscillation and t denotes the time. The velocity components u, v and w are assumed to be independent of y. The boundary layer and the continuity equations in a rotating frame of reference for a viscous fluid of density ϱ and kinematic viscosity ν are

$$\frac{\partial u}{\partial t} + u\frac{\partial u}{\partial x} + w\frac{\partial u}{\partial z} - 2\Omega v = -\frac{1}{\varrho}\frac{\partial p}{\partial x} + \nu\frac{\partial^2 u}{\partial z^2}, \tag{2.1}$$

$$\frac{\partial v}{\partial t} + u\frac{\partial v}{\partial x} + w\frac{\partial v}{\partial z} + 2\Omega u = -\frac{1}{\varrho}\frac{\partial p}{\partial y} + \nu\frac{\partial^2 v}{\partial z^2}, \tag{2.2}$$

$$\frac{\partial u}{\partial x} + \frac{\partial w}{\partial z} = 0. \tag{2.3}$$

The tangential pressure gradients generated by the oscillating outer flow are given by

$$-\frac{1}{\varrho}\frac{\partial p}{\partial x} = -U\omega \sin(kx + \omega t) - \frac{U^2 k}{2}\sin 2(kx + \omega t), \tag{2.4a}$$

$$-\frac{1}{\varrho}\frac{\partial p}{\partial y} = 2\Omega U \cos(kx + \omega t). \tag{2.4b}$$

These pressure gradients are in the travelling wave form and are out of phase with the free stream velocity. The tangential flow in the boundary layer is driven by the pressure gradients and the normal velocity arises out of conservation of mass. We examine the steady flow induced by the oscillatory external flow. Averaging the continuity equation (2.3) over one wave length, we obtain

$$\frac{d\bar{w}}{dz} = 0, \tag{2.5a}$$

where $\bar{w}$ is the steady normal velocity. The solution of (2.5a) satisfying the boundary condition

$$\bar{w} = 0 \quad \text{at} \quad z = 0 \tag{2.5b}$$

gives $\bar{w} = 0$ for all z. It follows that no mean normal velocity is produced by the oscillating external field. On the other hand steady velocity components in the x and y directions are anticipated. In order to be able to obtain the expressions for such a steady flow, the governing equations (2.1) to (2.3) are non-dimensionalized by the transformation

$$u' = \frac{u}{U}, \quad v' = \frac{v}{U}, \quad w' = \sqrt{\frac{\Omega}{\nu}}\frac{w}{kU}, \tag{2.6}$$

$$x' = kx, \quad z' = \sqrt{\frac{\Omega}{\nu}}\, z, \quad t' = \Omega t.$$

The resulting equations, after using (2.6) and dropping the primes, are

$$\frac{\partial u}{\partial t} + Ro\left(u\frac{\partial u}{\partial x} + w\frac{\partial u}{\partial z}\right) - 2v$$

$$= -\beta \sin(x + \beta t) - \frac{1}{2}\sin 2(x + \beta t) + \frac{\partial^2 u}{\partial z^2}, \tag{2.7}$$

$$\frac{\partial v}{\partial t} + Ro\left(u\frac{\partial v}{\partial x} + w\frac{\partial v}{\partial z}\right) + 2u = 2\cos(x + \beta t) + \frac{\partial^2 v}{\partial z^2}, \tag{2.8}$$

$$\frac{\partial u}{\partial x} + \frac{\partial w}{\partial z} = 0. \tag{2.9}$$

The non-dimensional constants present in the equations (2.7) and (2.8) are the Rossby number Ro and a frequency parameter β defined by $Ro = \dfrac{Uk}{\Omega}$ and $\beta = \dfrac{\omega}{\Omega}$. The relevant boundary conditions for the problem are

$$u = v = w = 0 \quad \text{at} \quad z = 0,$$

$$u \to \cos(x + \beta t), \quad v \to 0 \quad \text{as} \quad z \to \infty. \tag{2.10}$$

When the Coriolis force is large compared to the inertial force, the Rossby number is small.

In the following sections we obtain the expressions for the oscillating and steady velocity components when rotation is large. To this end, we assume, $Ro < 1$.

3. The Oscillating Flow

When $Ro < 1$, the velocity components u, v, w can be taken in the form

$$(u, v, w) = \sum_{n=0}^{\infty} Ro^n (u_n, v_n, w_n). \tag{3.1}$$

Substituting (3.1) in (2.7) to (2.10) and considering the zeroth order terms, we obtain

$$\frac{\partial u_0}{\partial t} - 2v_0 = -\beta \sin(x + \beta t) + \frac{\partial^2 u_0}{\partial z^2}, \tag{3.2}$$

$$\frac{\partial v_0}{\partial t} + 2u_0 = 2\cos(x + \beta t) + \frac{\partial^2 v_0}{\partial z^2}, \tag{3.3}$$

$$\frac{\partial u_0}{\partial x} + \frac{\partial w_0}{\partial z} = 0, \tag{3.4}$$

with

$$u_0 = v_0 = w_0 = 0 \quad \text{at} \quad z = 0, \qquad u_0 \to \cos(x + \beta t), \quad v_0 \to 0 \quad \text{as} \quad z \to \infty. \tag{3.5}$$

Owing to the harmonically oscillating pressure gradients and the free stream, an oscillatory flow alone can be expected in this order. For the solution of the system (3.2) and (3.3) subjected to (3.5), we set

$$q_0 = u_0 + iv_0 = q_{01}\, e^{i(x+\beta t)} + q_{-01}\, e^{-i(x+\beta t)}. \tag{3.6}$$

Substituting (3.6) in (3.2), (3.3) and (3.5) and solving we obtain

$$q_0 = \cos(x + \beta t) - \frac{1}{2} e^{-\sqrt{\frac{2+\beta}{2}}z} \exp\left[i\left(x + \beta t - \sqrt{\frac{2+\beta}{2}}\,z\right)\right] +$$

$$+ \frac{1}{2} e^{-\sqrt{\frac{2-\beta}{2}}z} \exp\left[-i\left(x + \beta t + \sqrt{\frac{2-\beta}{2}}\,z\right)\right], \tag{3.7}$$

and hence

$$u_0 = \cos(x + \beta t) - \frac{1}{2} e^{-\sqrt{\frac{2+\beta}{2}}z} \cos\left(x + \beta t - \sqrt{\frac{2+\beta}{2}}\,z\right) -$$

$$- \frac{1}{2} e^{-\sqrt{\frac{2-\beta}{2}}z} \cos\left(x + \beta t + \sqrt{\frac{2-\beta}{2}}\,z\right), \tag{3.8}$$

$$v_0 = -\frac{1}{2} e^{-\sqrt{\frac{2+\beta}{2}}z} \sin\left(x + \beta t - \sqrt{\frac{2+\beta}{2}}\,z\right) +$$

$$+ \frac{1}{2} e^{-\sqrt{\frac{2-\beta}{2}}z} \sin\left(x + \beta t + \sqrt{\frac{2-\beta}{2}}\,z\right). \tag{3.9}$$

From the conservation of mass, we obtain

$$w_0 = z \sin(x + \beta t) - \frac{1}{2\sqrt{2(2+\beta)}} e^{-\sqrt{\frac{2+\beta}{2}}z} \times$$

$$\times \left[\cos\left(x + \beta t - \sqrt{\frac{2+\beta}{2}}\,z\right) - \sin\left(x + \beta t - \sqrt{\frac{2+\beta}{2}}\,z\right)\right] +$$

$$+ \frac{1}{2\sqrt{2}}\left(\frac{1}{\sqrt{2+\beta}} - \frac{1}{\sqrt{2-\beta}}\right)\cos(x + \beta t) +$$

$$+ \frac{1}{2\sqrt{2(2-\beta)}} e^{-\sqrt{\frac{2-\beta}{2}}z} \left[\cos\left(x + \beta t + \sqrt{\frac{2-\beta}{2}}\,z\right) + \right.$$

$$+ \sin\left(x + \beta t + \sqrt{\frac{2-\beta}{2}}\,z\right)\bigg] -$$

$$- \frac{1}{2\sqrt{2}}\left(\frac{1}{\sqrt{2+\beta}} + \frac{1}{\sqrt{2-\beta}}\right)\sin(x + \beta t)\,. \tag{3.10}$$

We notice that the travelling wave oscillations of the external stream in the x direction are transmitted to the x and z components of the motion. The cyclonic and anti-cyclonic components of the imposed harmonic oscillations interact with the rotation which results in the bifurcation of the EKMAN layer into two layers with the depths of penetration of order $\left(\frac{\nu}{2\Omega \pm \omega}\right)^{1/2}$. The vorticity which pervades these layers is carried from the boundary by the exponentially decaying oblique diffusive waves travelling in the xz-plane. These waves travel in two definite directions and have the wave lengths

$$2\pi\left(\frac{2\nu k^2 + 2\Omega \pm \omega}{2\nu}\right)^{1/2}.$$

The phase lead and lag of the diffusive wave system depend on z. The phase differences depending on z in the velocity components u_0, v_0, w_0 are necessary to support a mean shear in the fluid and to produce velocity correlations in the fluid. The non-zero correlations of the velocity components, by way of inertial force, induce a steady state response in the higher orders. The first term in w_0 is linear in z and hence the solution is valid in the boundary layers near the boundary. At the first look, it may appear that a resonance phenomenon is possible when the forcing frequency ω is equal to the natural frequency 2Ω of the rotating fluid, i.e. when $\beta = 2$. In the analysis of the flow due to harmonic oscillations of the bounding plane in a rotating fluid, THORNLEY [6] found that the tangential velocity components do not satisfy the boundary conditions when $\omega = 2\Omega$. On the contrary, the solution is well behaved for this critical frequency when the plate is at rest and the external flow is temporally oscillating, THORNLEY [6]. It is to be noted that there is no normal flow in the study of THORNLEY [6], whereas in the present study, oscillating axial flow exists when the free stream is temporally as well as spatially oscillating. By evaluating the indeterminate limit $\beta \to 2$, we find that u_0, v_0 and w_0 satisfy all the boundary conditions. Thus resonant situation does not arise for the critical frequency.

4. The Steady Flow

The first order equations obtained by substituting (3.1) in (2.7) to (2.10) are

$$\frac{\partial u_1}{\partial t} - 2v_1 = -\frac{1}{2}\sin 2(x + \beta t) + \frac{\partial^2 u_1}{\partial z^2} - \left(u_0\frac{\partial u_0}{\partial x} + w_0\frac{\partial u_0}{\partial z}\right), \tag{4.1}$$

$$\frac{\partial v_1}{\partial t} + 2u_1 = \frac{\partial^2 v_1}{\partial z^2} - \left(u_0\frac{\partial v_0}{\partial x} + w_0\frac{\partial v_0}{\partial z}\right), \tag{4.2}$$

$$\frac{\partial u_1}{\partial x} + \frac{\partial w_1}{\partial z} = 0\,, \tag{4.3}$$

with

$$\begin{aligned} &u_1 = v_1 = w_1 = 0 \quad \text{at} \quad z = 0\,,\\ &u_1 \to 0\,, \qquad v_1 \to 0 \quad \text{as} \quad z \to \infty\,. \end{aligned} \tag{4.4}$$

From the equations (4.1) and (4.2) we observe that in the first order, the velocity field is due to the tangential pressure gradients and the non-zero inertial forcing from the zeroth order velocity correlations. Substitution of (3.8) to (3.10) in (4.1) and (4.2) indicates the structure of u_1 and v_1 in the form

$$q_1 = u_1 + iv_1 = q_{10} + q_{12}\,e^{2i(x+\beta t)} + q_{-12}\,e^{-2i(x+\beta t)}\,. \tag{4.5}$$

Introducing (4.5) in the system of equations (4.1) to (4.4) and solving the resulting differential equations with appropriate boundary conditions, we obtain

$$\begin{aligned} q_{10} = \bigg[&\frac{(1+i)}{4\beta} + \frac{1}{2\beta\sqrt{4-\beta^2}} - \frac{\beta}{8\sqrt{4-\beta^2}\,(1+\sqrt{4-\beta^2})} + \\ &+ \frac{\beta}{4(16-\beta^2)} - \frac{1}{\beta} + \frac{[(3+\beta) - i(1+\beta)]}{16[(2+\beta)^2+1]} - \\ &- \frac{[(3-\beta) - i(1-\beta)]}{16[(2-\beta)^2+1]}\bigg]\,e^{-(1+i)z} - \\ &- \frac{1}{8\beta}\left[1 + i + \sqrt{\frac{2+\beta}{2-\beta}} - \frac{4(2+\beta)}{\beta}\right]e^{-\sqrt{i(2+\beta)}\,z} - \\ &- \frac{1}{8\beta}\left[1 + i + \sqrt{\frac{2-\beta}{2+\beta}} + \frac{4(2-\beta)}{\beta}\right]e^{-\sqrt{i(2-\beta)}\,z} + \\ &+ \frac{e^{-\sqrt{-i(2+\beta)}\,z}}{8(4+\beta)} - \frac{e^{-\sqrt{-i(2-\beta)}\,z}}{8(4-\beta)} + \\ &+ \frac{[i(1+\beta) - (3+\beta)]}{16[(2+\beta)^2+1]}\,e^{-\sqrt{2(2+\beta)}\,z} + \\ &+ \frac{[(3-\beta) - i(1-\beta)]}{16[(2-\beta)^2+1]}\,e^{-\sqrt{2(2-\beta)}\,z} + \\ &+ \frac{\beta}{8\sqrt{4-\beta^2}\,(1+\sqrt{4-\beta^2})}\,e^{-\left(\sqrt{i(2+\beta)}+\sqrt{i(2-\beta)}\right)z} + \\ &+ \frac{\sqrt{i(2-\beta)}\,z}{4\beta}\,e^{-\sqrt{i(2-\beta)}\,z} + \frac{\sqrt{i(2+\beta)}\,z}{4\beta}\,e^{-\sqrt{i(2+\beta)}\,z}\,, \end{aligned} \tag{4.6}$$

$$\begin{aligned} q_{12} = \Bigg[&\left(\frac{2+\beta}{2\beta^2} - \frac{1}{4\beta} - \frac{1}{8(4+\beta)} + \frac{i}{8\beta}\sqrt{\frac{2+\beta}{2-\beta}} - \right. \\ &- \frac{\left\{(1+\beta) + i\left(\sqrt{4-\beta^2} + \sqrt{\frac{2+\beta}{2-\beta}}\right)\right\}}{16\sqrt{5-\beta^2}}\Bigg]\,e^{-\sqrt{2i(1+\beta)}\,z} + \\ &+ \left(\frac{1}{4\beta} - \frac{i}{8\beta}\sqrt{\frac{2+\beta}{2-\beta}} - \frac{2+\beta}{2\beta^2}\right)e^{-\sqrt{i(2+\beta)}\,z} + \\ &+ \frac{\left[(1+\beta) + i\left(\sqrt{4-\beta^2} + \sqrt{\frac{2+\beta}{2-\beta}}\right)\right]}{16(5-\beta^2)} \times \\ &\times e^{-\left(\sqrt{i(2+\beta)}+\sqrt{-i(2-\beta)}\right)z} + \\ &+ \frac{\sqrt{i(2+\beta)}\,z}{4\beta}\,e^{-\sqrt{i(2+\beta)}\,z} + \frac{e^{-\sqrt{-i(2-\beta)}\,z}}{8(4+\beta)}\,, \end{aligned} \tag{4.7}$$

$$\begin{aligned} q_{-12} = \Bigg[&\frac{1}{8(4-\beta)} - \frac{1}{4\beta} - \frac{(2-\beta)}{2\beta^2} + \frac{i}{8\beta}\sqrt{\frac{2-\beta}{2+\beta}} + \\ &+ \frac{\left\{(1-\beta) + i\left(\sqrt{4-\beta^2} + \sqrt{\frac{2-\beta}{2+\beta}}\right)\right\}}{16(5-\beta^2)}\Bigg]\,e^{-\sqrt{2i(1-\beta)}\,z} + \\ &+ \left(\frac{1}{4\beta} + \frac{(2-\beta)}{2\beta^2} - \frac{i}{8\beta}\sqrt{\frac{2-\beta}{2+\beta}}\right)e^{-\sqrt{i(2-\beta)}\,z} + \\ &+ \frac{\left[(1-\beta) + i\left(\sqrt{4-\beta^2} + \sqrt{\frac{2-\beta}{2+\beta}}\right)\right]}{16(5-\beta^2)} \times \\ &\times e^{-\left(\sqrt{i(2-\beta)}+\sqrt{-i(2+\beta)}\right)z} - \\ &- \frac{e^{-\sqrt{-i(2+\beta)}\,z}}{8(4-\beta)} + \frac{\sqrt{i(2-\beta)}\,z}{4\beta}\,e^{-\sqrt{i(2-\beta)}\,z}\,. \end{aligned} \tag{4.8}$$

The fluctuating normal velocity can be easily obtained from

$$w_{12} = -2i\int_0^z (q_{12} + \bar{q}_{12})\,dz\,, \tag{4.9a}$$

$$w_{-12} = 2i\int_0^z (q_{-1} + \bar{q}_{-12})\,dz\,, \tag{4.9b}$$

where

$$w = w_{12}\,e^{2i(x+\beta t)} + w_{-12}\,e^{-2i(x+\beta t)} \tag{4.9c}$$

and an overbar denotes the complex conjugate of the function under it.

From the structure of the mean velocity function q_{10}, it can be seen that an Ekman Layer of thickness of order $\left(\frac{\nu}{\Omega}\right)^{1/2}$ and another layer of thickness of order $\left[\left(\frac{\nu}{2\Omega+\omega}\right)^{1/2}+\left(\frac{\nu}{2\Omega-\omega}\right)^{1/2}\right]$ are present, apart from the boundary layers present in the zeroth order flow. The Coriolis, viscous and the driving mean shear forces strike a balance in these layers. The solution for $\beta = 2, 4$ do not present any difficulties. Hence no resonance effect appears even in this order for the cirtical frequencies.

Acknowledgement

The authors are thankful to Prof. P. S. Mani Sundaram, Principal, Regional Engineering College, Tiruchirapalli and Dr. V. P. Muthuswamy, Professor of Mathematics, Regional Engineering College, Tiruchirapalli for their unabiding interest in this work.

References

1 Stewartson, K., Advances in Applied Mechanics, VI, 1, Academic Press, New York (1960).
2 Stuart, J. T., Laminar Boundary Layers, 349, Clarendon Press, Oxford, England (1963).
3 Riley, N., Unsteady laminar boundary layers, SIAM Review **17**, 274 (1975).
4 McCroskey, W. J., Some Current Research in Unsteady Fluid Dynamics The 1976 Freeman Scholar Lecture, Trans. ASME Jl. Fluids Engg. **99**, 8 (1977).
5 Telionis, D. P., Review – Unsteady boundary layers separated and attached Trans. ASME. – Jl. Fluids Engg. **29**, 101 (1979).
6 Thornley, C., On Stokes and Rayleigh Layers in a rotating system, Q. J. Math. Appl. Mech. **21**, 451 (1968).
7 Patel, M. H., On laminary boundary layers in oscillatory flow, Proc. Roy. Soc. Lond. A **347**, 99 (1975).

Eingereicht am 22. 8. 1980

Anschrift: Mrs. Parvathy Rajagopalan and Dr. R. Purushothaman, Department of Mathematics, Regional Engineering College, Tiruchirapalli-620015, India

BUCHBESPRECHUNGEN

Leipholz, H., Stability of Elastic Systems. Mechanics of Elastic Stability 7. Alphen aan den Rijn, Sijthoff & Noordhoff Intern. Publ. 1980. 492 S., Dfl. 150.00. $ 75.00. ISBN 90-286-0050-7

Mit der „Stabilität elastischer Systeme“ legt der bekannte Autor ein neues Buch zur Stabilitätstheorie vor, das in 4 Abschnitte geteilt ist: Grundlagen einer Stabilitätstheorie, Lösungsmethoden, Klassen von Stabilitätsproblemen, Ausgewählte nichtkonservative Stabilitätsprobleme.

In den „Grundlagen“ werden nach einer kurzen Einführung in die Problematik des Stabilitätsbegriffes Methoden zur Aufstellung der Bewegungs-Differentialgleichungen und der Randbedingungen entwickelt. Durch die Einführung einer „Variationsableitung“ der potentiellen Energie gelingt es, Systeme mit endlich und mit unendlich vielen Freiheitsgraden relativ einheitlich zu behandeln und später im 2. Abschnitt die Ljapunovsche direkte Methode auf die Kontinua zu erweitern. Es wird ein Grundaxiom zur Stabilität eingeführt, mit dem sowohl Durchschlag- als auch Verzweigungsprobleme behandelt werden können. Der Abschnitt „Lösungsmethoden“ ist vor allem der Darlegung der Ljapunovschen direkten und der indirekten Methode sowie ihrer (schon erwähnten) Erweiterung auf Probleme mit unendlich vielen Freiheitsgraden gewidmet. Bei der Begründung der indirekten Methode bleibt allerdings die Frage offen, welchen Sinn die Einführung eines nicht positiv definiten Abstandsmaßes im Parameterraum hat (S. 87). Auch fällt es nicht immer leicht, die Abgrenzung zum 3. Abschnitt einzusehen, nicht alle der in diesem Buch häufigen Stoffwiederholungen dienen dem besseren Verständnis. Hervorzuheben sind die Untersuchungen über den Einfluß von Imperfektionen und zu stochastischen Systemen. Der letzte Abschnitt befaß sich vor allem mit der näheren Untersuchung des Einflusses von „follower forces“, d. h. von Kräften, deren Richtung an die Deformation des elastischen Systems gebunden ist. Für ausgewählte Systeme, vor allem am Stab, aber auch an Platten und Schalen werden die möglichen Instabilitätserscheinungen „divergence“ (statisches Versagen) und „flutter“ (Flattern) ausführlich untersucht.

Das Buch wird dem Lernenden wegen seiner gründlichen Darstellung des nicht einfach zu verstehenden Sachverhaltes willkommen sein, der Erfahrene wird manche neue Anregung finden. Wenn es auch nicht das Ziel des Buches ist, praktische Berechnungsverfahren für reale technische Systeme vorzustellen, so sind nach Meinung des Rezensenten die Anknüpfungspunkte dafür zu kurz gekommen. Auch wird vom Standpunkt der Praxis her bedauert, daß physikalische Nichtlinearitäten von vornherein ausgeschlossen wurden.

Magdeburg U. Fischer

Szczepiński, W., Introduction to the Mechanics of Plastic Forming of Metals. Mechanics of Plastic Solids 3. Alphen aan den Rijn, Sijthoff & Noordhoff Intern. Publishers 1979. 378 S., Dfl. 120.00. $ 60.00. ISBN 90 286 0126 0

Das Buch gibt eine systematische Darstellung der mathematischen Methoden der Plastizitätstheorie. Gegenstand sind Berechnungsmethoden im Rahmen der phänomenologischen Theorie. Die vorliegende englische Ausgabe basiert auf einer polnischen Ausgabe aus dem Jahre 1976.

Der Inhalt des Buches läßt sich drei Schwerpunkten zuordnen. Zunächst werden mit den Kapiteln 1 bis 3 die Grundlagen der technischen Plastomechanik bereitgestellt. Das betrifft insbesondere die plastischen Eigenschaften der Metalle, die Formänderungsgeschwindigkeiten, Fließbedingungen (Huber-Mises, Tresca), Materialverfestigung, Extremalprinzipe der Plastizität und die Gleichungen des dreidimensionalen plastischen Fließens. In den Kapiteln 4 bis 9 werden fundamentale Aufgaben der Plastizitätstheorie behandelt. Sie umfassen das plastische Fließen, das Eindrücken eines Stempels in einen Halbraum, die Plastifizierung eines Blockes zwischen zwei Stempeln, verschiedene Probleme des Stauchens, Schneidens, Ziehens, Tiefziehens, Fließpressens und Durchschlagens, stationäre und instationäre Vorgänge, das axialsymmetrische plastische Fließen sowie spezielle Fragen bei Vorliegen eines ebenen Spannungszustandes. Als Lösungsmethode für die praktisch besonders interessierenden ebenen und axialsymmetrischen Probleme werden die Gleitlinientheorie und Hodographennetze genutzt. Abschließend befassen sich die anwendungsorientierten Kapitel 10 und 11 mit Formprozessen bei Schalen. Vorgestellt wird das Fließpressen von Rohren unter Beachtung der nichtstationären, stationären und Übergangsphasen sowie das Ziehen und Streckziehen von dünnwandigen beliebig doppelt gekrümmten Schalen. Dabei werden die Charakteristiken der Differentialgleichungssysteme und die Erscheinungen beim Tiefziehen für verschiedene Fließbedingungen verglichen sowie der Einfluß der Reibung abgeschätzt.

Die behandelten Probleme sind von grundlegender Bedeutung für die Berechnung zahlreicher Vorgänge in der Umformtechnik und für einige Trennprozesse. Numerische Beispiele zu den theoretischen Lösungen zeigen die Relevanz der dargelegten Methoden für die Praxis. Zahlreiche Tabellen und Diagramme erleichtern die Benutzung der Verfahren in der Ingenieurpraxis. An einigen Stellen wird auch der experimentelle Nachweis für die Zweckmäßigkeit der entwickelten Modelle erbracht. Es ist darauf hinzuweisen, daß der Verfasser auf die Darlegung von Näherungslösungen vollständig verzichtet und sich auf die mechanisch-mathematischen Probleme der Thematik konzentriert. In sprachlicher Hinsicht könnte die Ausgabe verbessert werden.

Das Buch gibt eine gelungene Einführung in die Problematik und macht mit neuesten Fortschritten auf dem Gebiet der Plastizitätstheorie bekannt. Es ist daher für Studenten, Ingenieure und Wissenschaftler des Maschinenbaus zu empfehlen.

Berlin W. Kaiser

Akkas, N. (Ed.), Progress in Biomechanics. Proceedings of the NATO Advanced Study Institute, Ankara, July 1978. Series E: Applied Science 32. Alphen aan den Rijn, Sijthoff & Noordhoff 1979. 395 S., Dfl. 75.00. $ 37.50. ISBN 90-286-0479-0

Ein einleitender Artikel dieses Sammelbandes befaßt sich mit Fragen der inneren Struktur und Führung heterogen zusammengesetzter Arbeitsgruppen, wie sie für interdisziplinäre Zusammenarbeit typisch sind, z. B. gerade auf dem Feld der Biomechanik, aber nicht nur dort. Der nächste Beitrag beschäftigt sich mit dem Verhältnis von Medizin und Bioengineering bei der Rehabilitation. Die neun restlichen Arbeiten lassen sich vier Themenkreisen zuordnen, die zu den meistuntersuchten der Biomechanik gehören: 1. Verhalten weicher Gewebe, vor allem der Haut; Ausblick auf die plastische Chirurgie; 2. Gang- und Gelenkanalyse; Ausblick auf die Prothetik; 3. Untersuchungen an Knochen: Wellenausbreitung; Mikrostruktur und Fraktur; 4. Modellierung des Schädels, Kopf- und Genickverletzungen. Die Beiträge wollen mehr einen allgemeinen Ein- und Überblick bzw. Literaturbericht geben als Detailfragen abhandeln.

Berlin W. Heinrich

Hirschel, E. H. (Ed.), Proceedings of the Third GAMM-Conference on Numerical Methods in Fluid Mechanics. Notes on Numerical Fluid Mechanics 2. DVFLR, Köln, October 1979. Braunschweig, Vieweg 1980. VIII, 315 S., DM 98,–. ISBN 3-528-08076-0

In diesem Buch werden 31 Vorträge der 3. GAMM-Konferenz, die vom 10. bis 12. Oktober 1979 in Köln stattfand, veröffentlicht. Die Mehrzahl der Beiträge befaßt sich mit der Anwendung von numerischen Methoden zur Lösung der Grundgleichungen der Gas- und Hydrodynamik bei konkreten Strömungsproblemen. Einige Arbeiten gehen auch auf Diffusions-, Wärmeleitungs- und Stabilitätsprobleme ein oder untersuchen laminare und turbulente Grenzschichten. Bei den numerischen Methoden dominieren Differenzenverfahren und Methoden der finiten Elemente. Nur in wenigen Beiträgen werden Spektral- und Kollokationsverfahren angewandt oder die numerischen Methoden selbst zum Gegenstand der Untersuchung genommen. Vorwiegend werden zweidimensionale Aufgabenstellungen gelöst, bei denen die Gleichungen sowohl für die Geschwindigkeit und den Druck als auch für die Stromfunktion und die Wirbelgeschwindigkeit formuliert werden. Bei komplizierteren Strömungsgebieten werden vor ihrer Diskretisation Transformationen auf einfachere eingeführt. Die erzielten numerischen Ergebnisse sind zum Teil durch den Einsatz von Plottern übersichtlich dargestellt.

Das Buch gibt auch einen Einblick in die letzten Entwicklungen der numerischen und rechentechnischen Methoden bei der Lösung von ingenieur-technischen Aufgaben aus den Gebieten des Strömungsmaschinen- und Energieanlagenbaus.

Berlin W. Zwick

Rath, H. J., Peristaltische Strömungen. Medizinische Informatik und Statistik 19. Berlin-Heidelberg-New York, Springer-Verlag 1980. VIII, 119 S., DM 25,–. US $ 14.80. ISBN 3-540-09999-9

In der vorgelegten kleinen Monografie wird eine Übersicht über die Nachbildung peristaltischer Strömungen im Experiment sowie über ihre mathematische Modellierung und über die Güte der Modelle gegeben. Zur Orientierung des Lesers wird diese Darstellung durch eine Erklärung der Phänomene der Peristaltik in der Physiologie und durch die Erörterung der theoretischen Grundlagen ergänzt. Für den Harnleiter, den Dünndarm und den Samenleiter werden recht einfache Modelle angegeben, die von der Annahme einer schleichenden Strömung und eines Newtonschen Fluids ausgehen und spezielle Forderungen an die Wellen stellen. Die Beurteilung der Qualität dieser Modelle fällt schwer, da gesichertes experimentelles Material kaum vorliegt. Ein nichtlineares Modell für eine allgemeine peristaltische Strömung mit sinusförmiger Wellenbewegung liefert für Potentialströmung numerische Lösungen, die mit dem Experiment gut übereinstimmen. Weitere Modelle, die, wie die anderen Modelle auch, der Literatur entnommen wurden, ermöglichen die Einbeziehung der Wandeigenschaften (Viskoelastizität, Wandrauhigkeit o. ä.), anderer Fließgesetze als dem Newtonschen Gesetz usw. Die Darstellung des Stoffes beschränkt sich insgesamt auf das Wesen der Modelle, ohne detailliert auf die Durcharbeitung und Lösung des jeweiligen Problems einzugehen. Man erhält zwar so einen gedrängten Überblick über das Vorhandene, ist aber auf die Originalliteratur verwiesen, wenn Einzelheiten interessieren. Die für den Mediziner geschriebene Publikation kann dem Mathematiker oder dem Mechaniker nur dann Anregungen geben, wenn er sich erstmalig über den Gegenstand informieren will.

Berlin K. Remke

Śniatycki, J., Geometric Quantization and Quantum Mechanics. Applied Mathematical Sciences 30. Berlin-Heidelberg-New York, Springer-Verlag 1980. IX, 230 S., DM 28,–. US $ 16.60. ISBN 3-540-90469-7

Unter „geometrischer Quantisierung" darf man einen Zweig der Differentialgeometrie verstehen, der dem physikalisch-heuristischen Begriff der „Quantisierung" eines klassischen Systems eine genaue mathematische Bedeutung beizulegen versucht. Insbesondere handelt es sich darum, wie und ob man von gegebenen Poissonklammern (oder, was dasselbe ist, von einer geschlossenen simplektischen Form) und gegebener Hamiltonfunktion zu Kommutatoren entsprechender Operatoren eines Hilbertraums gelangt. Die hierzu notwendigen zusätzlichen Vorschriften werden dabei in geometrischen Begriffen ausgedrückt. Der Verfasser, J. Śniatycki, beschreibt kurz aber erfreulich klar die mathematische Struktur der „geometrischen Quantisierung", um sodann an wesentlichen Anwendungsbeispielen darzustellen, in welchem Umfang man durch diese Theorie tatsächlich auf die Quantenmechanik mit Notwendigkeit trifft. Natürlich zeigen alle Erfahrungen, daß es keinen eindeutigen Weg von der klassischen zur Quantenmechanik gibt. Das Buch hinterläßt den Eindruck, daß das allgemeine Vorgehen und das Begriffssystem der Theorie gut gewählt sind, im Einzelnen die verschiedenen Anwendungsfälle jedoch Modifikationen verlangen, die offenbar diese Nichteindeutigkeit widerspiegeln und die, wenigstens zur Zeit, keiner Begründung zugängig sind. Das Buch beginnt mit einer interessanten Einleitung, in der alle späteren Kapitel vorbesprochen werden.

Hier finden sich auch historische Bemerkungen und Hinweise auf ungelöste und auf unverstandene Probleme. In den Kapiteln 2 bis 6 gibt der Autor eine Einführung in die „geometrische Quantisierung“. Dann folgen als Anwendungen ein Kapitel über SCHRÖDINGERdarstellungen eines und mehrerer Teilchen und den Zusammenhang zur quasiklassischen Näherung. Hiernach werden die FOCK-BARGMANN-Darstellung erörtert und die zeitabhängige SCHRÖDINGERgleichung aus der Theorie abgeleitet. Als Beispiele für relativistische Dynamiken folgen: Relativistische Teilchen im elektromagnetischen Feld, die Ladungsüberauswahlregeln und die Quantisierung in der KALUZA-Theorie. Das letzte Kapitel ist dem Problem gewidmet, wie man in der „geometrischen Quantisierung“ mit dem Spin auskommt.

Das Buch zeigt, daß es besonders in den letzten Jahren gelungen ist, auch anspruchsvollere Anwendungen der Theorie der geometrischen Quantisierung zu erschließen. Doch ist ihr Hauptfeld wohl die mathematische Analyse von Zusammenhängen zwischen Klassischer und Quantenmechanik. Das Buch trägt zum tieferen Verständnis dieser Zusammenhänge bei und kann mit Gewinn gelesen werden.

Leipzig A. UHLMANN

Heymann, M., Structure and Realization Problems in the Theory of Dynamical Systems, Wien-New York, Springer-Verlag 1975. 79 S., S 117,–. DM 17,– (CISM Courses and Lectures 204)

Die moderne Theorie der linearen dynamischen Systeme geht auf die fundamentalen Arbeiten von R. E. KALMAN zurück. Sie wurde vor allem durch die Anforderungen aus den Gebieten der optimalen Steuerung und Schätzung von Prozessen initiiert, hat aber auch weitreichende Konsequenzen auf andere wichtige Gebiete der Dynamik. Insbesondere der enge Zusammenhang zwischen Steuerbarkeit und Beobachtbarkeit führt zu intensiven Studien über die innere Struktur linearer Systeme.

Die vorliegende Abhandlung ist der Strukturtheorie und einigen Realisierungsproblemen bei linearen Systemen gewidmet. Neben den für das Verständnis notwendigen allgemein bekannten Resultaten über Steuerbarkeit, Beobachtbarkeit, Dekomposition enthält das Buch vor allen Dingen zur Bruchdarstellung, zur Realisierung und zu rückgekoppelten Systemen neue Aussagen. Der Inhalt ist unterteilt in: I. Allgemeine Begriffe linearer dynamischer Systeme; II. Erreichbarkeit, Beobachtbarkeit, Minimalität; III. Struktur des Zustandsraumes; IV. Äquivalenztheorie linearer Übertragungssysteme; V. Bruchdarstellung und Rückkopplung.

Der Stoff wird in knapper und kompakter Weise dargeboten. Sein Studium setzt Vorkenntnisse auf dem Gebiet der linearen Systemtheorie voraus (etwa wie in: ATHANS, M., FALB, P. L., Optimal Control, McGraw, New York, 1966). Das Literaturverzeichnis enthält 4 Zitate und stellt eine geeignete Auswahl über moderne Entwicklungslinien der linearen Systemtheorie dar.

Berlin B. HEIMANN

Hawking, S. W. / Israel, W. (Eds.), General Relativity. An Einstein Centenary Survey. Cambridge, Cambridge University Press 1980. XVIII, 919 S., £ 12.50 A H/C. ISBN 0-521-29928-4

Das EINSTEIN-Centenarium 1979 hat zu einer reichen internationalen literarischen Aktivität Anlaß gegeben, wobei einerseits die biographische EINSTEIN-Literatur stark vermehrt und andererseits der physikalischen Fundierung der Physik durch EINSTEIN auf den Gebieten der Relativitätstheorie, Quantentheorie und Statistik durch moderne Beiträge besondere Aufmerksamkeit geschenkt wurde. Der von den beiden renommierten Relativitätstheoretikern HAWKING und ISRAEL herausgegebene Sammelband widmet sich ausschließlich der Relativitätstheorie. Er umfaßt ein sehr reiches Spektrum von der experimentellen Verifizierung der EINSTEINschen Gravitationstheorie über die weitere mathematische Durchdringung dieser Theorie bis hin zur relativistischen Astrophysik und Kosmologie sowie Quanten-Gravitation. Die wichtigsten Gegenstände seien in folgendem durch Stichworte kurz erwähnt: WILL, Äquivalenzprinzip, post-Newtonsche Gravitation im Sonnensystem, Binärpulsar-System, Einfluß der Globalstruktur des Universums auf die lokale Physik; DOUGLASS und BRAGINSKY, Quellen und Detektoren von Gravitationswellen; FISCHER und MARSDEN, Anfangswertproblem und dynamische Formulierung der EINSTEIN-Theorie; GEROCH und HOROWITZ, Globalstruktur der Raum-Zeiten; CARTER, Mechanische, elektromagnetische und thermodynamische Eigenschaften von Schwarzen Löchern; CHANDRASEKHAR, Theorie der KERR-Metrik und deren Störungen; BLANDFORD und THORNE, Astrophysik der Schwarzen Löcher; DICKE und PEEBLES, Urknall-Kosmologie; ZEL'DOVICH, Kosmologie des frühen Universums; MacCALLUM, Anisotrope und inhomogene relativistische Weltmodelle; PENROSE, Singularitäten und Zeit-Asymmetrie; GIBBONS, Quanten-Feldtheorie in der gekrümmten Raum-Zeit; DeWITT, Quanten-Gravitationen als neue Synthese; HAWKING, Der Path-Integral-Zugang zur Quanten-Gravitation; WEINBERG, Ultraviolett-Divergenzen in der Quantentheorie der Gravitation.

Der Auswahl dieser international herausragenden Autoren und der heute aktuellsten Forschungsgegenstände ist es zu verdanken, daß dieser Sammelband ein überdurchschnittlich hohes Niveau besitzt. Der Fachmann wird umfassend über die neuesten Erkenntnisse informiert. Das Werk ist für den Forscher auf dem Gebiet der Relativitätstheorie ein unentbehrliches Arbeitsmaterial und kann wärmstens empfohlen werden. Einen besonderen Wert gewinnt es noch darüberhinaus dadurch, daß auf mehr als 70 Seiten eine aktualisierte Bibliographie zusammengefaßt ist.

Jena E. SCHMUTZER

Heinhold, J. / Behringer, F., Einführung in die Höhere Mathematik. Teil 3: Differentialgleichungen. München-Wien, Carl Hanser Verlag 1979. 432 S., 48 Abb., DM 44,–. ISBN 3-446-12316-4

An den Band 2 der „Einführung in die Höhere Mathematik“, die der Differential- und Integralrechnung gewidmet ist (Referat in ZAMM 57 (1977), S. 562) schließt sich der vorliegende dritte Band über Differentialgleichungen an. Mit einem vierten Band über Funktionentheorie wird das Werk seinen Abschluß finden.

Im ersten Kapitel werden nach allgemeinen Begriffsbildungen einfache Typen expliziter und impliziter gewöhnlicher Differentialgleichungen erster Ordnung eingeführt. Danach werden für gewöhnliche Differentialgleichungen höherer Ordnung die Herabsetzung der Ordnung und die Zurückführung auf Systeme von Differentialgleichungen erster Ordnung sowie umgekehrt die Zurückführung eines Systems von Differentialgleichungen erster Ordnung auf eine Differentialgleichung höherer Ordnung behandelt. Damit können die anschließenden Existenz- und Eindeutigkeitsaussagen auf Systeme von Differentialgleichungen erster Ordnung beschränkt werden. Einer genaueren Betrachtung werden lineare Differentialgleichungen erster und höherer Ordnung unterzogen, wobei unter anderem das als Resonanzkatastrophe bezeichnete unbeschränkte Wachsen der Schwingungsamplitude bei erzwungenen harmonischen Schwingungen erhalten und mit der EULERschen, der BESSELschen, der LAGUERREschen, der HERMITEschen, der LEGENDREschen, der GAUSSschen und der KUMMERschen Differentialgleichung auf Beispiele für lineare Differentialgleichungen mit nicht konstanten Koeffizenten kurz eingegangen wird.

Ein besonderes Kapitel ist der Lösung von Anfangs-, Rand- und Eigenwertproblemen auf Analog- und Digitalrechnern gewidmet. Einführende Betrachtungen, Durchrechnung einfacher Beispiele mit verschiedenen Methoden, wobei das RUNGE-KUTTA-Verfahren, die „PICARD“-Iteration und die Prediktor-Korrektor-Verfahren den Schwerpunkt bilden, und weiterführende Hinweise etwa zur Schrittweitensteuerung oder zum Einsatz von Hybridrechnern bilden eine gelungene Einheit.

Berlin G. SCHMIDT

Heinhold, J. / Gaede, K.-W., Einführung in die Höhere Mathematik. Teil 4: Funktionentheorie. München-Wien, Carl Hanser Verlag 1980. XII, 548 S., ISBN 3-446-12317-2

Mit dem vorliegenden Teil 4 findet die „Einführung in die Höhere Mathematik“ ihren Abschluß (Teil 1: Grundlagen der

Linearen Algebra; Teil 2: Infinitesimalrechnung, beide 1976, ZAMM 57 (1977), S. 562; Teil 3: Differentialgleichungen, 1975, Rezension voranstehend). Er ist wie die anderen Teile aus Vorlesungen und Übungen für Studenten der Physik und der Elektrotechnik der ersten vier Semester hervorgegangen. Das Charakteristikum dieser Darstellung, die vielfältige Vorknüpfung der grundlegenden mathematischen Begriffe und Zusammenhänge wie holomorphe und meromorphe Funktionen, konforme Abbildungen, Integration holomorpher Funktionen, CAUCHYscher Integralsatz und CAUCHYsche Integralformel und SCHWARZsche Polygonabbildung mit Veranschaulichungen und Anwendungen in Strömungsmechanik und Elektrostatik sind nicht nur für Studenten dieser Anwendungsdisziplinen, sondern auch der Mathematik selbst von großem Nutzen. Nützlich und ausbauwürdig erscheinen auch die Hinweise auf den Einsatz von Digital- und Analogrechnern bei der konformen Abbildung, wobei freilich auf eine gesonderte Buchdarstellung von HEINHOLD und KULISCH verwiesen werden konnte. Eine eingehende Behandlung der LAPLACEtransformation mit Anwendungen zur Lösung von Differential- und Integralgleichungen und der Differentialgleichungen für komplexe Funktionen einer komplexen Veränderlichen rundet das Gesamtwerk ab. In diesem wie in den früheren Teilen ergänzt eine Vielzahl von Übungsaufgaben mit ausführlichen Lösungen den Text. So ist eine Gesamtdarstellung der Höheren Mathematik entstanden, die für Vorlesungen und zum Selbststudium gleichermaßen geeignet ist und einen festen Platz in der Lehrbuchliteratur einnehmen wird.

Berlin G. SCHMIDT

Campbell, S. L. / Meyer, C. D. jr., Generalized Inverses of Linear Transformations. London-San Francisco-Melbourne, Pitman 1979. XI, 272 S., £ 23.00. ISBN 0-273-08422-4

Bekanntlich sind die verallgemeinerten Inversen zu einem überaus leistungsfähigen Werkzeug bei der Lösung linearer mathematischer Probleme herangereift. Vielfältige Anwendungen in den verschiedensten mathematischen Gebieten und in einer Reihe von naturwissenschaftlichen und ingenieurwissenschaftlichen Gebieten unterstreichen ihre praktische Bedeutung. Seit der Veröffentlichung der beiden grundlegenden Arbeiten des Engländers PENROSE aus den Jahren 1955 und 1956 ist eine fast unüberschaubare Flut von Beiträgen zu diesem Gebiet veröffentlicht worden (Ref. schätzt etwa 2500). Bis 1974 erschien etwa ein halbes Dutzend Monographien oder Lehrbücher über verallgemeinerte Inversen von Matrizen und 1976 eine umfangreiche, kommentierte Bibliographie von NASHED und RALL mit 1776 Literaturstellen, die versuchte, den damals erreichten Stand in Theorie und Anwendung zusammenzufassen.

Das hier zu besprechende Buch wendet sich, wie die Autoren betonen, gleichermaßen an die in der Forschung Tätigen, an Praktiker und an Studenten. In der Tat überspannt das Buch die ganze Breite von den schon klassischen Resultaten bis zu neuen Forschungsergebnissen. Es behandelt eine Reihe von Anwendungen, z. B. Parameterschätzungen nach der Methode der kleinsten Quadrate, Theorie elektrischer Netzwerke, Theorie endlicher MARKOFFscher Ketten, Theorie linearer Systeme von Differentialgleichungen, lineare Optimierung. Und um den Studenten entgegenzukommen, werden nur die Elemente der linearen Algebra vorausgesetzt, werden die Beweise ausführlich dargestellt, so daß sie leicht nachvollzogen werden können.

Die Autoren hatten sich das Ziel gestellt, ein leicht lesbares, einführendes Buch zu schreiben, das zum weiteren Lernen anregen soll. Ich glaube, dieses Ziel wurde erreicht. An vielen Stellen, wo der aufmerksame Leser Fragen stellen würde, finden sich schon die erläuternden Antworten. Verschiedentlich eingestreute interessante Bemerkungen lockern den Text auf. Erwähnt sei hier nur die nette Geschichte der Bahnberechnung der Ceres durch GAUSS und die dadurch ermöglichte Wiederauffindung dieses kleinen Planeten. Eine vereinfachte Version dieses Problems wird durch Verwendung von verallgemeinerten Inversen gelöst.

Zur Charakterisierung des Inhalts seien die Kapitelüberschriften angeführt: 0. Introduction and other preliminaries, 1. The MOORE-PENROSE or generalized inverse, 2. Least squares solutions, 3. Sums partitioned matrices and the constraint generalized inverse, 4. Partial isometrics and EP matrices, 5. The generalized inverse in electrical engineering, 6. (i, j, k)-Generalized inverses and linear estimation, 7. The DRAZIN inverse, 8. Applications of the DRAZIN inverse to the theory of finite MARKOV chains, 9. Application of the DRAZIN inverse, 10. Continuity of the generalized inverse, 11. Linear programming, 12. Computational concerns.

Die ausführliche Behandlung der sog. DRAZIN-Inversen, die bei der Lösung linearer Differentialgleichungssysteme eine immer wichtigere Rolle spielt, nimmt in diesem Buch eine zentrale Stelle ein. Dies ist auch der wichtigste Unterschied zu früheren Büchern. Das Buch enthält 180 Übungsaufgaben, im allgemeinen ohne Lösung. Mit Literaturangaben wird relativ sparsam umgegangen, es wird auf die oben erwähnte Bibliographie verwiesen. Dies ist legal, führt aber z. B. dazu, daß der relativ hohe Anteil des Buches an eigenen Ergebnissen der Autoren ziemlich verborgen bleibt und der Anschluß zu weiterer Forschung nicht erleichtert wird.

Insgesamt ist das Buch sehr sorgfältig geschrieben, gut lesbar und übersichtlich gedruckt und kann allen empfohlen werden, die sich in das interessante Gebiet einarbeiten wollen, die verallgemeinerte inverse Matrizen anwenden oder sich in der Forschung näher damit befassen wollen.

Halle G. ZIELKE

Brezinski, C., Padé-Type Approximation and General Orthogonal Polynomials. ISNM 50. Basel-Boston-Stuttgart, Birkhäuser Verlag 1980. 250 S., sFr. 62.—. DM 68,—. US $ 34.00. ISBN 3-7643-1100-2

Eine Padé-Approximierende (PA) ist dadurch definiert, daß ihre Potenzreihe eine vorgegebene Potenzreihe so weit wie möglich approximiert. Durch diese Eigenschaft sind sowohl der Zähler als auch der Nenner einer PA eindeutig bestimmt. Problematisch ist hierbei die Lokalisierung der Polstellen. Man hat keine Kontrolle über die Polstellen und kann sie nicht in bestimmten Bereichen der komplexen Ebene ansiedeln. Aus diesem Grund betrachtet der Autor Padé-Typ-Approximierende (PTA), bei denen es möglich ist, einige der Pole oder auch alle frei zu wählen. Der Zähler und der noch nicht bestimmte Nenneranteil werden dadurch definiert, daß eine vorgegebene Potenzreihe soweit wie möglich zu approximieren ist.

Ein Ziel des Buches ist, PTA einzuführen. Das geschieht im ersten Kapitel. Hervorzuheben sind hier die vom Autor aufgezeigten offenen Probleme und auch die Hinweise auf Anwendungen. PTA sind sehr nützlich in den Anwendungen, da mit ihnen bessere Resultate als mit PA erzielt werden können. Die Definition der PTA erlaubt auch, auf die Wahl von Polen zu verzichten. Die PTA ist dann die PA. Das zweite Ziel dieses Buches ist, PA über die Definition der PTA zu behandeln. Dies führt zur Einführung von allgemeinen orthogonalen Polynomen in die Theorie der PA.

In Kapitel II werden die wesentlichsten Aspekte der allgemeinen orthogonalen Polynome zusammengestellt.

Kapitel III ist den PA gewidmet. Der Zugang über die PTA in Verbindung mit den allgemeinen orthogonalen Polynomen führt zu einer geschlossenen algebraischen Theorie der PA. Alte und neue Resultate werden gewonnen, wie rekursive Schemata zur Berechnung von PA, Fehler-Formeln, Matrix-Interpretation und Konvergenz beschleunigende Methoden für Folgen von Zahlen und Vektoren.

Ein viertes Kapitel beinhaltet Verallgemeinerungen von PTA.

Das Buch ist sehr anregend, da der angebotene Stoff zu vielen offenen Problemen, insbesondere auf dem Gebiet der PTA führt.

Berlin W. ZIELAS

Plachky, D. / Baringhaus, L. / Schmitz, N., Stochastik I, Eine elementare Einführung in Grundbegriffe der Wahrscheinlichkeitsrechnung und Statistik. Studienbuch für Studierende der Mathematik, Natur- und Wirtschaftswissenschaften ab 2. Semester. Wiesbaden, Akademische Verlagsgesellschaft 1978. 250 S., DM 22,—. ISBN 3-400-00390-5

Die Autoren formulieren als Ziel dieses Bandes „Stochastik I", Studierende der Mathematik mit Vorkenntnissen aus Anfängervorlesungen „elementar und systematisch" in die stochastische Denkweise einzuführen. Eine maßtheoretische Grundlegung der Stochastik, die über das Modell einer diskreten Wahrscheinlichkeitsverteilung hinausgeht, wird in einem Band „Stochastik II" behandelt. Das gestellte Ziel ist auf engstem Raum beispielhaft erreicht. Dabei erscheinen Vorkenntnisse in mathematischer Denkweise und mathematischen Beweismethoden in der Tat unerläßlich. Was dafür durch Konzentration auf das Wesentliche und durch meisterhafte Verknüpfung geeigneter Beispiele mit voll bewiesenen theoretischen Zusammenhängen alles aufgenommen ist, ist erstaunlich. Nach einer Einführung in die Kombinatorik werden LAPLACE-Verteilungen und allgemeiner endliche und abzählbar unendliche diskrete Wahrscheinlichkeitsverteilungen, bedingte Wahrscheinlichkeiten und unabhängige Ergebnisse behandelt. Den Hauptteil bildet die Theorie der Zufallsgrößen, ihrer Verteilungen, Verteilungsfunktionen, Erwartungswerte und Streuungen, ihrer stochastischen Konvergenz und Unabhängigkeit, der erzeugenden Funktionen ganzzahliger Zufallsgrößen sowie bedingter Verteilungen und bedingter Erwartungswerte. Als Anwendung wird auf 34 Seiten eine Einführung in die Schätztheorie gegeben, in der auf die Erwartungstreue (bis zur Bestimmung gleichmäßig bester erwartungstreuer Schätzungen), Suffizienz (Kriterium von NEYMAN) und Vollständigkeit eingegangen wird.

Berlin G. SCHMIDT

Feichtinger, G., Demographische Analyse und populationsdynamische Modelle. Grundzüge der Bevölkerungsmathematik. Wien-New York, Springer-Verlag 1979. XVI, 367 S., 45 Abb., S 690,–. DM 96,–. US $ 53.80. ISBN 3-211-81540-6

Das Ziel seiner Ausführungen, so der Autor, sei erreicht, „wenn dem Leser die Notwendigkeit der Analyse und des Modellbaues bei der Lösung demographischer Fragen einsichtig wird". Dies ist sehr bescheiden formuliert, besagte Notwendigkeit wird wohl nirgendwo mehr bestritten, wo die Thematik überhaupt reflektiert wird, das gestellte Ziel wird also sicher erreicht, spätestens beim Blättern in dem Buch: Ist es vielleicht nicht interessant, daß noch 80% der ledigen 25-jährigen Frauen eine künftige Heirat erwarten können, aber nur noch 50% der 32-jährigen? Oder was der Unterschied zwischen Querschnitts- und longitudinaler Betrachtungsweise in der Bevölkerungsmathematik bzw. was Kohortenprozesse sind? Zur Sache: Das Buch bietet eine bisher fehlende sehr gut lesbare Einführung in den Sprachgebrauch, die Methodik und die Denkweise der formalen Demographie. Einführung heißt hier, daß die Beschränkungen bei der Stoffauswahl didaktischen Gesichtspunkten folgen, so werden zum Beispiel nur abgeschlossene Bevölkerungen behandelt, keine Migrationseffekte; formal meint, daß etwa soziologische, ökonomische, kulturelle Mechanismen außer Diskussion bleiben. Es handelt sich also nicht um ein Handbuch für fertige Spezialisten, sondern um ein Lehrbuch für Statistiker, Sozial- und Wirtschaftswissenschaftler, relevant für Planung und Verwaltung. Die Beschreibung dessen, was das Buch nicht ist, soll nichts Negatives sagen, sondern vielmehr das Anliegen und die Vorzüge des Buches herausstreichen. Die zur Illustration verwendeten Daten beziehen sich überwiegend auf Oesterreich und Westdeutschland. Ein großer Teil des Buches ist mit relativ geringen mathematischen Vorkenntnissen aus Analysis und Wahrscheinlichkeitsrechnung lesbar.

Berlin W. HEINRICH

Fletcher, R., Unconstrained Optimization. Practical Methods of Optimization 1. Chichester-New York-Brisbane-Toronto, John Wiley & Sons 1980. VIII, 120 S., £ 8.80. ISBN 0-471-27711-8

In zunehmendem Maße werden numerische Verfahren zur Lösung von Optimierungsproblemen in Naturwissenschaft, Technik und Ökonomie eingesetzt. Sowohl zur Formulierung von Optimierungsaufgaben als auch zu deren sachgemäßer Behandlung z. B. mit einem der in Programmbibliotheken angebotenen Algorithmen erweisen sich Grundkenntnisse über Optimierungsaufgaben als erforderlich. Mit dem vorliegenden Buch gibt Professor FLETCHER, der durch eine große Zahl originaler Arbeiten die Entwicklung der Optimierungsverfahren selbst mitbestimmte, eine übersichtliche Darstellung zu den Verfahren zur unrestringierten Minimierung. Geschickt konzentriert sich der Autor dabei auf wesentliche Eigenschaften der Algorithmen, gibt geometrisch einfache Interpretationen an und vermeidet eine zu große Allgemeinheit. Auf diese Weise erhält der Leser einen guten Überblick über die Verfahren zur Optimierung ohne Restriktionen, wird mit den prinzipiellen Arbeitsweisen unterschiedlicher Verfahrensklassen vertraut gemacht und kann sich exemplarisch in einige Beweistechniken einarbeiten.

Das Buch gliedert sich mit Einleitung, Struktur der Methoden, NEWTON-ähnliche Verfahren, Verfahren der kunjugierten Richtungen, Methoden mit beschränkten Schrittweiten, Summe von Quadranten und nichtlineare Gleichungen in sechs Kapitel. Die Darstellung ist klar und verständlich gehalten und eignet sich ebensogut als Grundlage für Vorlesungen wie zum Selbststudium. Übungsaufgaben fördern das Verständnis des dargebotenen Stoffes. Der interessierte Leser findet zahlreiche Hinweise zur praktischen Realisierung und wird mit über 100 Zitaten an das Studium der Originalarbeiten herangeführt, wobei sich der Autor leider ausschließlich auf den eigenen Sprachraum beschränkt.

Insgesamt läßt sich erwarten, daß das vorliegende Buch einen breiten Kreis von Interessenten finden wird, und man kann schon erwartungsvoll dem Erscheinen des angekündigten zweiten Bandes entgegensehen.

Dresden Ch. GROSSMANN

Dixon, L. C. W. / Szegö, G. P. (Eds.), Numerical Optimisation of Dynamic Systems. Amsterdam-New York-Oxford, North-Holland Publ. Co. 1980. VIII, 412 S., US $ 53.75. Dfl. 110.00. ISBN 0-444-85494-0

Dieses Buch ist ein Sammelband von 22 Beiträgen zur Optimierung dynamischer Systeme und dazu in mehr oder weniger engen Beziehungen stehenden allgemeinen Grundlagen der Optimierungstheorie. Es umfaßt neben einem verbindenden Übersichtsartikel der Herausgeber weitere 9 weitgehend voneinander unabhängige Artikel zur dynamischen Optimierung, 7 Beiträge zur nichtlinearen Optimierung und 5 Arbeiten zur Sensitivitätsanalyse. Darstellung und Diskussion numerischer Verfahren der Optimierung in spezifizierten Formen für stetige und diskrete Systeme nehmen dabei einen breiten Raum ein, auch zahlreiche neuere konkrete Anwendungen kommen dabei zu Wort. Beweistechnische Aspekte werden hingegen zumeist unterdrückt unter Bezug auf die in der Regel recht reichhaltigen Literaturverzeichnisse. Auf diese Weise gleicht dieses Buch einem Tagungsbericht über neuere Entwicklungen zur Praxis und Theorie der dynamischen Optimierung. Es vermittelt Mathematikern wie Anwendern eine Fülle von Anregungen zum dargelegten Problemkreis. Die nachfolgende Übersicht der enthaltenen Beiträge mag diese Vielfalt verdeutlichen: 1. DIXON SZEGÖ, The Numerical Optimisation of Dynamic Systems: A Survey; 2. BARTHOLOMEW-BIGGS; The Optimisation of Spacecraft Orbital Manoeuvres Part I: Linearly Varying Thrust Angles; 3. BARTHOLOMEW-BIGGS, The Optimisation of Spacecraft Orbital Manoeuvres Part II: Using PONTRYAGIN's Maximum Principle; 4. OVERTON/TUNNICLIFFE WILSON, A Comparison of the Minimum Principle and Differential Programming; 5. CICHOCKI, Application of Augmented Lagrangians in Dynamic Planning Models; 6. PROPOI, Dynamic Linear Programming; 7. NURMINSKI, An Application of Nondifferentiable Optimization in Optimal Control; 8. CHICHINADZE, Solving Dynamic Optimal Problem by the ψ-Transformation Method; 9. WILSON, An Optimal Control Algorithm for Monotonically Structured Functions; 10. JAMES, A Recursive Variable Metric Method; 11. SCHITTKOWSKI, A Numerical Comparison of 13 Nonlinear Programming Codes with Randomly Generated Test Problems; 12. TAPIA, On the Role of Slack Variables in Quasi-NEWTON Methods for Nonlinear Optimization with Equality Constraints; 13. BERTOCCHI/CAVALLI/SPEDICATO, Computational Performance of Diagonalized Multiplier Quasi-NEWTON Methods for Nonlinear Optimization with Equality Constraints; 14. GOMULKA, Decomposition Algorithms for Smooth Unconstrained Optimisation Problems; 15. VUJCIC, Some Properties of Feasible Direction Methods; 16. ALUFFI/INCERTI/ZIRILLI, Systems of Equations and A-Stable Integration of Second

Order O.D.E.'s; 17. SUTTI, Superlinear Convergence of SUTTI's Non Gradient Minimization Algorithm; 18. FIACCO, Nonlinear Programming Sensitivity Analysis Results Using Strong Second Order Assumptions; 19. MCKEOWN, An Approach to Sensitivity Analysis; 20. ARCHETTI/BALLABLIO/VERCELLIS, Cost-Benefit Analysis of Insulation in Buildings Via Nonlinear Optimization; 21. ARCHETTI/BETRO, Optimization Problems Arising in the Design and Exploitation of Seismographic Networks; 22. BAIER, Mathematical Programming in Engineering Design Problems.

Leipzig R. KLÖTZLER

Todd, J., Basic Numerical Mathematics, Vol. 1: Numerical Analysis. ISNM 14. Basel-Stuttgart, Birkhäuser Verlag 1979. 253 S., sFr. 54,–. ISBN 3-7643-0729-3

Unter der Voraussetzung, daß dem Leser die Grundlagen der Differential- und Integralrechnung bereits bekannt sind, wird er mit Problemen und Gedankengängen vertraut gemacht, die für numerische Auswertungen typisch sind. Bei der iterativen Lösung von Gleichungen lernt er viel über Konvergenzgeschwindigkeit und Möglichkeiten zur Konvergenzbeschleunigung. Neben der gleichmäßigen Approximation wird auch die Bedeutung der asymptotischen Approximation für die Numerik herausgearbeitet. Die letzten drei Kapitel befassen sich exemplarisch mit der Interpolation, den darauf aufbauenden Quadraturverfahren sowie mit Differenzengleichungen und ihrer Anwendung bei der Lösung gewöhnlicher Differentialgleichungen, ohne daß eine abgerundete Theorie angestrebt wird. Ein Anhang über BESSELfunktionen behandelt die Grundlagen für die komplizierten Beispiele des vorhergehenden Textes.

Die auf Grund von Lehrerfahrungen in drei Jahrzehnten sorgfältig ausgewählten und erprobten Verfahren bieten zusammen mit der eingehenden Diskussion ihrer Anwendungsbereiche (Fehlerabschätzungen) sowie ihrer Anwendungsgrenzen (Instabilität) einen lebendigen Einblick in die Wirkungsweise numerischer Algorithmen im Zeitalter der modernen Rechentechnik.

Besonders anregend und lehrreich sind die 167 gestellten Übungsaufgaben und die dazu auf 68 Seiten ausführlich dargelegten Lösungshinweise. Die Palette der Aufgaben reicht dabei von solchen, die theoretische Einsichten vermitteln, bis zu Aufgaben, bei denen die Durchführung von Computerexperimenten geübt wird.

Unbefriedigend erscheint Ref. lediglich, daß das Symbol „$\sim$" nach seiner Definition für asymptotische Entwicklungen im Sinne von POINCARÉ (S. 75) später ohne genaue Definition auch für Funktionen mit Nullstellen (S. 77) sowie stillschweigend für asymptotische Darstellungen (S. 79, 111, 211) und außerdem in einer weiteren nicht näher präzisierten Bedeutung (S. 201) verwendet wird.

Rostock L. BERG

Jennings, A., Matrix Computation for Engineers and Scientists, London-New York-Sydney-Toronto, John Wiley and Sons, 1977, 330 S., £ 10.50. $ 22.00.

Für die Erstellung von effektiven Computerprogrammen, die den Matrizenkalkül zum Gegenstand haben, ist die alleinige Kenntnis der Matrizenalgebra nicht ausreichend. Dies trifft insbesondere dann zu, wenn man es — wie häufig in den Anwendungen — mit großen Matrizen zu tun hat. Genau diesem Problemkreis, also der Entwicklung von effektiven computergerechten Methoden, der Darstellung der Probleme, die bei der Umsetzung auf dem Rechner entstehen, ist das vorliegende Buch gewidmet.

In Kapitel 1 wird ein kurzer Abriß des Matrizenkalküls sowie der für Handrechnung geeigneten numerischen Verfahren gegeben. Es folgt ein Kapitel, in dem in einige ausgewählte Anwendungen des Matrizenkalküls eingeführt wird, die vornehmlich der Elektrotechnik und Mechanik entnommen sind. Im 3. Kapitel werden am Beispiel der Matrizenmultiplikation verschiedene effektive Speicherungstechniken behandelt, die besonders wichtig für große und wenig besetzte Matrizen sind. Die Kapitel 4, 5 und 6 haben die bekannten numerischen Methoden zur Auflösung linearer Gleichungssysteme zum Gegenstand, wobei das gesamte fünfte Kapitel der Darstellung effektiver Lösungstechniken für wenig besetzte Matrizen gewidmet ist. Das Kapitel 7 führt am Beispiel ausgewählter Anwendungen in das Eigenwertproblem ein, und die darauffolgenden drei Kapitel behandeln die entsprechenden numerischen Verfahren sowie ihre Verwirklichung auf dem Automaten (z. B. JACOBI-Verfahren, Tridiagonalisierung nach GIVENS, HANSHOLDER-Transformation, HESSENBERGverfahren in seiner automatengerechten Darstellung nach WILKINSON, LR-Transformation nach RUTISHAUSER, QR-Transformation nach FRANCIS sowie Vektoriterationsmethoden).

In den Kapiteln 2, 3, 4 und 5 sind für verschiedene Programmsegmente simultane ALGOL- und FORTRAN-Versionen aufgeführt. Diese sind weniger für eine direkte Computerbenutzung gedacht, vielmehr dienen sie zur Illustration der Programmierungstechniken. Das Buch dürfte sowohl bei Studenten als auch Ingenieuren eine sehr freundliche Aufnahme finden.

Berlin A. RENGER

Kazmierczak, H. (Hrsg.), Erfassung und maschinelle Verarbeitung von Bilddaten. Grundlagen und Anwendungen. Wien-New York, Springer-Verlag 1980. X, 399 S., 236 Abb., DM 168,–. S 1200,–. US $ 99.20. ISBN 3-211-81555-4

Mit diesem Buch wird der lohnenswerte Versuch unternommen, in einer Überschau das sich immer weiter auffächernde und ständig ausweitende Gebiet der automatischen Erfassung und interaktiven Verarbeitung von Bilddaten zusammenzufassen. Hierzu beschreiben mehrere Autoren aus unterschiedlichen Forschungseinrichtungen der BRD in Einzelbeiträgen Grundlagen und Resultate von Verarbeitungsverfahren sowie ihre Anwendungen, wie sie bis etwa 1979 auszugsweise aus der Literatur hervorgehen. — Zunächst werden Definitionen zur Bildverarbeitung und Erläuterungen zur Digitalisierung, Rasterung, Grauwertwiedergabe und Farbcodierung gegeben und die lokale Bildverarbeitung zur Objektverdünnung, Filterung, Faltung, Gradientenbildung sowie die globale Verarbeitung skizziert. Dabei wird auf die irrelevanz- und redundanzreduzierte Codierung zur Anpassung an das menschliche Auge eingegangen. Die Bildauswertung wird durch Rekonstruktion, Restaurierung, Detektion von veränderlichen Bildobjekten, Pixelklassifikation für Multispektralanalyse, Texturanalyse, Objektisolierung, Extraktion von Linienmustern, von Objektkonturen und von Oberflächen in Grauwertbildern erläutert. — Die Bildverarbeitenden Prozessorsysteme sind eingeteilt in a) vollständig parallelverarbeitende, aber kaum programmierbare köhärent- bzw. inköhärent-optische Prozessoren zur zweidimensionalen FOURIERtransformation, Faltung, Korrelation, Differentiation bzw. zur Erkennung und Orientierung von Objekten sowie in eingeschränkt programmierbare hybrid-optische Verfahren zur Faltung und Korrelation und in b) leicht programmierbare, aber aufwendige digitale Prozessoren in Form des VON-NEUMANN-Rechners (SI/SD-Typ), des Feldrechners (SI/MD-Typ), des Pipelining-Systems (MI/SD-Typ) und des Multiprozessorsystems (MI/MD-Typ), die sich durch spezielle Prozessorsysteme (Macro Arithmetic Processor) für schnelle FOURIER-Transformation, Faltung und Korrelation, für die binäre Bildverarbeitung (Cellular Logic Array for Image Processing) zur Konturliniendarstellung, Objektextraktion und morphometrisch-topologische Bildauswertung sowie durch flexiblere Prozessoren (Flexible Image Processor) für lokale Verarbeitung von Grauwertbildern ergänzen lassen. — Aus dem weiten Gebiet der Anwendung werden die Schriftzeichenerkennung, die Produktionskontrolle und Materialprüfung mit Texturauswertung und Pixelmanipulation, die Produktionsautomatisierung durch spezielle Sensoren und durch Manipulierende und navigierende Roboter, die Auswertung von Röntgen-, Mikroskop-, Ultraschall- und Thermographiebildern in der Medizin und in einem breiteren Rahmen die Gewinnung, Auswertung und Speicherung von Geodaten für die Erderkundung im thermischen, optischen und Mikrowellenbereich in ihren prinzipiellen Zügen dargestellt. Abschließend wird auf das sich entwickelnde Gebiet der Bildtransformationssysteme (Speichern, Sortieren, Verknüpfen und Manipulieren) für topographische und strukturelle Daten hingewiesen. — Die Themenfülle des Buches hat die Autoren zur Auswahl und Einschränkung gezwungen, die den Leser nicht immer zufriedenstellen. Eine einheitliche Darstellung des Stoffgebietes wäre wünschenswert, wenn damit Wiederholungen vermieden, stoffernere Gebiete beseitigt, eine kritische Wertung der behandelten Verfahren und eine gezieltere Literaturauswahl erreicht werden könnten. Dessenungeachtet ist das Buch jedem zu empfehlen, der sich zur Einarbeitung in die Bildverarbeitung einen allseitigen Überblick verschaffen will, während es dem Fachmann Hinweise auf angrenzende Arbeitsgebiete vermittelt.

Berlin U. RÖSLER

Hestenes, M., Conjugate Direction Methods in Optimization. Applications of Mathematics 12. Berlin-Heidelberg-New York, Springer-Verlag 1980. X, 325 S., 22 Abb., DM 59,50. US $ 35.10. ISBN 3-540-90455-7

Im Jahre 1952 erschien die Arbeit von M. R. HESTENES und E. STIEFEL über die Methode der konjugierten Gradienten zur Lösung linearer Gleichungssysteme, von der umfangreiche Aktivitäten mit dem Ergebnis ausgelöst worden sind, daß numerisch wirksame Algorithmen für speziell strukturierte Gleichungssysteme und Optimierungsaufgaben ohne und mit Restriktionen entwickelt werden konnten. Heutzutage gehören Verfahren dieser Gruppe zu den viel benutzten bei sparse Problemen, wie sie z. B. bei der Finitisierung von Randwertaufgaben entstehen. Die Aufnahme dieser Verfahren auch in einführende Darstellungen der numerischen Mathematik ist üblich geworden.

Das vorliegende Buch über die Methoden der konjugierten Richtungen, verfaßt von einem ihrer beiden Väter, verfolgt die Zielstellung, eine möglichst umfassende Darlegung der grundlegenden Ideen sowie wesentlicher Realisierungen zu geben. Eine vollständige Literaturauswertung ist nicht angestrebt worden. Im Mittelpunkt stehen Beschreibung und Analyse der Verfahren für Aufgaben mit quadratischer Zielfunktion und ohne Restriktionen sowie Ausdehnungen mit Hilfe modifizierter NEWTON-Verfahren auf den nichtquadratischen Fall. Im vorbereitenden ersten Kapitel werden Gradienten- und NEWTON-Verfahren im Hinblick auf Optimierungsaufgaben dargelegt. Die eigentliche Thematik des Buches beginnt mit dem zweiten Kapitel. Es ist der allgemeinen Methode der konjugierten Richtungen gewidmet, während das dritte und das vierte Kapitel mit dem konjugierten GRAM-SCHMIDT-Prozeß und der Methode der konjugierten Gradienten zwei der bedeutendsten Vertreter der allgemeinen Methode ausführlich behandeln. Der übersichtlich und dem Leser entgegenkommend abgefaßten, mit vielen numerischen Erfahrungen des Autors durchsetzten Monographie sollte ein breiter Nutzerkreis gewiß sein.

Dresden J. W. SCHMIDT

Mittelmann, H. D. / Weber, H. (Eds.), Bifurcation Problems and Their Numerical Solution. ISNM 54. Workshop, Dortmund, January 15—17, 1980. Basel-Boston-Stuttgart, Birkhäuser Verlag 1980. 243 S., sFr. 39,—. ISBN 3-7643-1204-1

Der vorliegende ISNM-Band enthält Vorträge, die auf einer internationalen Tagung über Bifurkationsprobleme und ihre numerische Lösung Anfang 1980 in Dortmund gehalten wurden. Die Beiträge reichen von neuen Resultaten auf dem Gebiet der LJUSTERNIK-SCHNIRELMAN-Theorie und der Wechselwirkung von Bifurkationsmoden bis hin zu numerischen Methoden zur Berechnung von Verzweigungspunkten für gewöhnliche und partielle Differentialgleichungen. Die Herausgeber haben an den Anfang ihres Berichtsbandes einen sehr informativen Übersichtsartikel über die verschiedenen Untersuchungsmethoden der Bifurkationstheorie gestellt. Im einzelnen enthält der Band folgende Beiträge: MITTELMANN, H. D. and WEBER, H., Numerical Methods for Bifurcation Problems — A Survey and Classification; BEYN, W.-J., On Discretizations of Bifurcation Problems; BONGERS, A., Über ein Rayleigh-Ritz-Verfahren zur Bestimmung kritischer Werte; KÜPPER, T., Pointwise Error Bounds for the Solutions of Nonlinear Boundary Value Problems; LANGFORD, W. F. and IOOSS, G., Interactions of HOPF and PITCHFORK Bifurcations; MOORE, G. and SPENCE, A., The Convergence of Approximations to Nonlinear Equations at Simple Turning Points; SCHOLZ, R., Computation of Turning Points of the Stationary NAVIER-STOKES Equations Using Mixed Finite Elements; SEYDEL, R., Programme zur numerischen Behandlung von Verzweigungsproblemen bei nichtlinearen Gleichungen und Differentialgleichungen; VOSS, H., Lower Bounds for Critical Parameters in Exothermic Reactions; WEBER, H., Shooting Methods for Bifurcation Problems in Ordinary Differential Equations; WERNER, B., Turning Points of Branches of Positive Solutions; WIESWEG, U., Eine numerische Behandlung von primären Bifurkationszweigen.

Berlin K. R. SCHNEIDER

Rautmann, R. (Ed.), Approximation Methods for Navier-Stokes Problems. Proceedings, Paderborn, Germany 1979. Lecture Notes in Mathematics 771. Berlin-Heidelberg-New York, Springer-Verlag 1980. XVI, 581 S., 202 Abb., DM 58,—. US $ 34.30. ISBN 3-540-09734-1

In diesem Band sind 34 Vorträge veröffentlicht, die auf dem IUTAM-Symposium über Approximationsmethoden für NAVIER-STOKES-Probleme von 9. bis 15. September 1979 in Paderborn gehalten wurden. Neben den NAVIER-STOKES-Gleichungen wird auch die BOUSSINESQgleichung behandelt. Bei den untersuchten Aufgaben überwiegen zweidimensionale laminare Strömungen inkompressibler Medien. Die meisten Arbeiten befassen sich mit Fragen der Approximation dieser Gleichungen durch Differenzen-, Spektral- und finite Elemente-Verfahren. Bei komplizierteren Strömungsvorgängen wird die GALERKIN-Variante der finiten Elemente-Methode bevorzugt. Zur numerischen Lösung der approximierenden Gleichungen werden verschiedene iterative Verfahren eingesetzt. Insbesondere soll auf die Realisierung eines Lösungsverfahrens mit Hilfe eines STOKES-Solvers hingewiesen werden. In zahlreichen Arbeiten werden auch Fragen der Lösbarkeit und Stabilität der Gleichungen sowie Fragen der Existenz, der Eindeutigkeit bzw. der Bifurkation, der Regularität und des asymptotischen Verhaltens der Lösungen untersucht. Dabei werden auch funktionalanalytische Methoden angewandt. Spezielle Probleme der Stabilität bzw. Instabilität der Gleichungen werden am Beispiel der RAYLEIGH-TAYLOR-Instabilität, der COUETTE-Strömung und der Strömung im Spalt zweier konzentrischer Kugeln gelöst. Bei der Berechnung von Strömungen werden sowohl Durchström- als auch Umströmprobleme, sowohl stationäre als auch instationäre Vorgänge betrachtet. Dabei wird neben der Geschwindigkeits- und Druckdarstellung auch von der Formulierung der Bewegungsgleichung mit Hilfe der Strömfunktion und Wirbelgeschwindigkeit ausgegangen. Viele Strömungsmechaniker und angewandte Mathematiker werden in diesem Band Anregungen zur Lösung ihrer Probleme finden.

Berlin W. ZWICK

Girault, V. / Raviart, P.-A., Finite Element Approximation of the Navier-Stokes Equations. Lecture Notes in Mathematics 749. Berlin-Heidelberg-New York, Springer-Verlag 1979. VII, 200 S., 4 Abb., DM 25,—. US $ 13.80. ISBN 3-540-09557-8

Diese aus Spezialvorlesungen entstandene Schrift handelt von den mathematischen Grundlagen zur approximativen Lösung der NAVIER-STOKESschen Gleichungen für inkompressible Strömungen. Zu diesem Zweck werden neben den Räumen, in denen die Lösungen liegen, endlich dimensionale Unterräume betrachtet. Die Darstellung beginnt mit den STOKESschen (linearisierten) Problemen, zunächst geht es um stationäre, dann um instationäre Strömungen. Es folgen die vollständigen nichtlinearen Probleme in der gleichen Reihenfolge.

Die fünf Kapitel haben die Überschriften: 1. Mathematische Grundlagen des STOKESschen Problems; 2. Numerische Lösung des STOKESschen Problems. Eine klassische Methode; 3. Eine gemischte Finite-Elemente-Methode zur Lösung des STOKESschen Problems; 4. Die stationären NAVIER STOKESschen Gleichungen; 5. Die instatonären NAVIER STOKESschen Gleichungen. Während in den ersten Kapiteln eine ausführliche lehrbuchartige Darstellung vorherrscht, wird der Stil in den letzten beiden Kapiteln gedrängter. Im 4. Kapitel wird auch die approximative Bestimmung einer nicht notwendig eindeutig bestimmten Lösung erörtert. Besonders interessant im letzten Kapitel ist die Semi-Diskretisierung mittels Mehrschrittmethoden, mit deren Hilfe eine quadratisch konvergente Folge von Näherungslösungen gewonnen wird. Am Schluß findet man zu jedem Kapitel bibliographische Bemerkungen und ein 47 Titel umfassendes Literaturverzeichnis.

Dies ist ein mathematisches Buch; es kann auch all den Anwendern empfohlen werden, die bei der numerischen Behandlung von Strömungsproblemen gern auf sicherem Boden stehen. Vieles von dem, was hier über die NAVIER-STOKESschen Gleichungen gesagt wurde, läßt sich auf andere nichtlineare Probleme übertragen.

Berlin J. FÖRSTE

Bauer, K. W. / Ruscheweyh, S., Differential Operators for Partial Differential Equations and Function Theoretic Applications. Lecture Notes in Mathematics 791. Berlin-Heidelberg-New York, Springer-Verlag 1980. V, 258 S., DM 29,—. US $ 17.20. ISBN 3-540-09975-1

Dieses Buch besteht aus den beiden Teilen „Differentialoperatoren für partielle Differentialgleichungen" (190 S.) von K. W. BAUER und „Über die Funktionentheorie der PESCHL-BAUER-Gleichung" (68 S.) von S. RUSCHEWEYH. Im ersten Teil wird ein Überblick gegeben über 113, im zweiten Teil über 68 Arbeiten zu den angeführten Themen, und es werden auch neue Ergebnisse eingeflochten.

Ausgangspunkt des Buches sind lineare, homogene Differentialgleichungen in der komplexen Form

$$W_{z\bar{z}} + A(z, \bar{z})\, w_{\bar{z}} + B(z, \bar{z})\, w = 0 ,$$

deren Lösungen in der Gestalt

$$w = \sum_{k=0}^{n} a_k(z, \bar{z}) \left(a(z) \frac{\partial}{\partial z}\right)^{z} g(z) + \sum_{k=0}^{m} b_k(z, \bar{Z}) \left(\overline{b(z)} \frac{\partial}{\partial \bar{z}}\right)^{k} \overline{h(z)}$$

mit beliebigen holomorphen Funktionen $g(z)$, $h(z)$ gesucht werden. Von besonderem Interesse ist der Fall, daß man auf diesem Wege die allgemeine Lösung erhält. Später werden die Funktionen g, h durch Elemente allgemeinerer Funktionenklassen (Lösungen anderer partieller Differentialgleichungen) ersetzt, und nach geeigneten Transformationen treten auch Integrale in den Lösungsdarstellungen auf. Die Ergebnisse werden zum Teil auf inhomogene Gleichungen, auf Gleichungen höherer Ordnung sowie auf den Fall mehrerer unabhängiger Veränderlicher übertragen, und es wird auch ein Zusammenhang zu den Lösungsdarstellungen von BERGMANN bzw. VEKUA hergestellt.

Ein ganzes Kapitel ist den Anwendungen auf Spezielle Funktionen und ihre Differentialgleichungen gewidmet, die in verschiedenen Gebieten der Physik und Technik auftreten, wie Kugelflächenfunktionen, pseudoanalytische Funktionen, Tricomi Gleichung, STOKES-BELTRAMI-Systeme, Gleichung der axialsymmetrischen Potentialtheorie und Verallgemeinerungen dieser Gleichungen.

Der zweite Teil befaßt sich mit dem weitergehenden Studium der speziellen Gleichungen

$$(1 + \varepsilon z\bar{z})^2\, w_{z\bar{z}} + \varepsilon n(n + 1)\, w = 0$$

mit $\varepsilon = \pm 1$ und natürlichen Zahlen n. Der BAUERsche Darstellungssatz ermöglicht den Aufbau einer zugehörigen Funktionentheorie, die durch die folgenden Kapitelüberschriften gekennzeichnet ist: Struktur der Lösungen; DIRICHLETprobleme für den Kreis; Funktionen mit eingeschränktem Wertebereich, SCHWARZsches Lemma; univalente Lösungen, RIEMANNscher Abbildungssatz; Räume vom HARDY-Typ; Summierbarkeit, ABELscher Satz; Probleme der Werteannahme; Eindeutigkeitssätze; isolierte Singularitäten, PICARDscher Satz; analytische Fortsetzung; automorphe Funktionen.

Da hiermit die Möglichkeiten zur Übertragung von Ergebnissen der klassischen Funktionentheorie keineswegs erschöpft sind, werden die Leser zur Mitarbeit bei der Schließung der Lücken aufgerufen.

Rostock L. BERG

Sydow, A. (Ed.), Systems Analysis and Simulation 1980. Proceedings of the Internat. Symposium, Berlin (GDR), September 1—5, 1980. Berlin, Akademie-Verlag 1980. 496 S., M 62,—. BN 762-811-0 (2182/5)

The book contains about 250 contributions. The main contributions (invited papers) are given from outstanding specialists: 1. N. S. RAJBMAN: Dispersion identification of nonlinear systems. Within a literature survey the author defines dispersion functions, considers their properties and discusses dispersion equations of identification and identifiability. 2. A. SYDOW: Mathematical and simulation models in systems analysis. The author reports on research done at the ZKI concerning the analysis. of real existing systems and the methodology of analysis and simulation of systems. 3. K. H. FASOL: Theoretical model building and simulation of industrial processes. The basic ideas and the general principles of process analysis and modelling are outlined. An example for model order reduction is discussed. 4. E. J. H. KERCKHOFFS: Parallel simulation — A survey of local research. The current research at Delft University of Technology is reported. Attention is paid to applications in architecture and in the Delft time-shared minisystem. 5. K. BELLMANN, K. STEINMÜLLER: Systems analysis and modeling in ecology. Some existing simulation models, intrinsinc features of ecosystems, problems of control and management of ecosystems are discussed. 6. G. C. VANSTEENKISTE: Simulation of environmental systems: an academic exercise? The rapid expansion of research into the modelling of complex environmental systems is pointed out to be not notably successful in practical terms. The reasons are investigated. 7. M. PESCHEL: Multiobjective decision making in systems analysis. The concept of Pareto-optimality is used to tackle multiobjective optimization problems. Applications are presented (designing textile articles, operating a city heating system). 8. I. TROCH: Optimization and other non-classical concepts in system modelling. The use of optimal control theory, catastrophe theory, and the theory of dynamic systems is demonstrated for modelling of "soft" systems. 9. G. SCHWARZE: Towards software for modelling and simulation. A survey on the present state is given and simulation systems developed in the GDR are outlined. 10. K. REINISCH, J. WERNSTEDT, A. BECKER: Contribution of methods of systems analysis and controlling large-scale systems and their application to operating water resources.

Most of the other contributions comprise only one page. The cover the following subjects: 1. Methods of system analysis, 2. Model building and identification, 3. Simulation systems, 4. Applications to industrial and technological systems (Chemical reactors, Energy systems, Technological processes, Thermical systems), 5. Application to traffic systems, 6. Applications to environmental, ecological and biological systems (especially agriculture systems, Aquatic ecosystems, Water systems, Theory of ecological systems), 7. Application to global and regional systems (Energy problems, Social systems analysis).

Thus the book can be recommended to both the theoretician and the practician.

Dresden F. BEICHELT

Helgason, S., The Radon Transform. Progress in Mathematics 5. Boston-Basel-Stuttgart, Birkhäuser Verlag 1980. VII, 192 S., sFr. 15.—. ISBN 3-7643-3006-6

Mit dieser Schrift liegt ein Lehrbuch über Theorie und Anwendung der RADON-Transformation vor. Es handelt sich allgemein um das Problem der Bestimmung von Funktionswerten aus Integralwerten über bestimmte Mannigfaltigkeiten. Im Kap. 1. wird die Aufgabe behandelt, im n-dimensionalen euklidischen Raum aus gegebenen Integralwerten über d-dimensionale Ebenen die Funktion selbst zu bestimmen. Nach Definition der RADON-Transformation und der zugehörigen dualen Transformation werden mit Hilfe der RIESZschen Integraloperatoren Umkehrformeln bewiesen. Anwendungsbeispiele sind das CAUCHY-Problem der Wellengleichung und die Durchleuchtungs-Inversion. Im 2. Kap. wird die Theorie verallgemeinert auf lokalkompakte topologische Gruppen. Das 3. Kap. behandelt den Fall einer RIEMANNschen Mannigfaltigkeit. Hier wird die RADON-Transformation definiert durch Integrale über geodätische Mannigfaltigkeiten. Für Räume konstanter Krümmung wird ein Umkehrtheorem bewiesen. Im 4. Kap. werden pseudo-RIEMANNsche Mannigfaltigkeiten mit indefiniter Metrik betrachtet. Ein verallgemeinertes RIESZ-Potential wird eingeführt, das in einfachem Zusammenhang mit dem Wellenoperator steht. Es gilt ein Umkehrtheorem. Die Schrift enthält ferner den Orginalabdruck der Arbeit von J. RADON aus dem Jahre 1917. Mit dem Lesen dieses Buches gewinnt man einen guten Überblick nicht nur über die allgemeine Theorie, sondern auch über die Möglichkeiten der Anwendung.

Berlin E. HEYN

Frehse, J. / Pallaschke, D. / Trottenberg, U. (Eds.), Special Topics of Applied Mathematics. Functional Analysis, Numerical Analysis and Optimization. Proceedings of the Seminar in Bonn, 8 — 10 October, 1979. Amsterdam-New York, North-Holland Publ. 1980. VIII, 242 S., US $ 39.00. Dfl. 80.00. ISBN 0-444-86035-5

Der Band enthält Beiträge zu drei Themenkreisen von Teilnehmern des Seminars, welches aus Anlaß der Emeritierung von Prof. H. Unger abgehalten wurde. Entsprechend seiner eigenen Arbeitsrichtung nennt der Rezensent bezüglich des ersten und dritten Gebietes nur die Namen der Autoren: H. Jarchow, W. B. Johnson, D. Preworska-Rolewicz, D. Pumplün, J. B. Retherford, W. Ruess, Zb. Semanedi; D. Hausmann, B. Korte, S. Rolewicz, A. Bacopoulos, G. Godini, I. Singer.

Im Teil über Numerische Analysis fallen sofort die zwei Beiträge von A. Brandt und W. Hackbusch ins Auge, da sie nahezu dem gleichen Thema gewidmet sind, der Mehrgittermethode zur Lösung von Randwertaufgaben für partielle Differentialgleichungen. Sie ergänzen sich aber in schöner Weise: A. Brandt geht mehr auf Fragen der Implementierung im Zusammenhang mit FEM-Verfahren ein, W. Hackbusch auf die Konvergenz. P. Deuflhard und V. Apostolescu untersuchen verschiedene Aspekte der Gauss-Newton-Methode zur Lösung nichtlinearer Regressionsprobleme: Konvergenz, Abänderung des Abstiegskriteriums, Schrittweitenwahl, numerische Realisierung, und zeigen einige numerische Ergebnisse. C. Temperton berichtet von Varianten der weiteren Beschleunigung der komplexen und der reellen numerischen Fourier-Transformation, vor allem auch im Hinblick auf die Möglichkeiten von Vektorcomputern. H. Werner betrachtet nichtlineare Splines bei der Lösung gewöhnlicher Differentialgleichungen (darunter mit Singularitäten), gibt dazu Konvergenzergebnisse und eine Anwendung auf Epidemie-Modelle an. Der Artikel von A. Brandt ist der längste in diesem Teil, dennoch aber sehr dicht geschrieben und von erstaunlicher, durch vielseitige Anwendung des Dargelegten geprägten Gründlichkeit.

Obwohl bei allen diesen Artikeln schnell klar wird, von welchen Veröffentlichungen hier zusammenfassende, kommentierende, verkürzende oder erweiternde Darstellungen gegeben werden, sind sie interessant und enthalten originelles, lehrreiches Material, das von dem theoretischen Niveau und der aus ernsthaften Anwendungen in anderen Wissensgebieten erworbenen praktischen Erfahrungen gleichermaßen Zeugnis gibt.

Berlin G. Stoyan

Beckenbach, E. F. (Ed.), General Inequalities 2. Proceedings of the Second Intern. Conference, Oberwolfach 1978. ISNM 47. Basel-Boston-Stuttgart, Birkhäuser Verlag 1980. XX, 485 S., sFr. 60,—. ISBN 3-7643-1056-1

Die zweite internationale Tagung über Ungleichungen in Oberwolfach stand unter der Leitung von J. Aczél (Waterloo, Ont.), E. F. Beckenbach (Los Angeles), R. Ger (Katowice) und W. Walter (Karsruhe). Der jetzt vorliegende Tagungsbericht enthält alle Arbeiten, die auf der Tagung vorgestellt worden sind, und darüber hinaus Beiträge, die vorgesehen waren, jedoch nicht gehalten werden konnten. Die etwa drei Dutzend Arbeiten haben in der Regel den Charakter von Zeitschriftenveröffentlichungen. Sie überstreichen mit ihrer Thematik ziemlich viele Gebiete und sind vom Herausgeber zu folgenden Komplexen zusammengefaßt worden: Ungleichungen in Kombinatorik und Zahlentheorie, Ungleichungen in der Matrizentheorie und linearen Algebra, Ungleichungen bei Differential- und Integraloperatoren, Ungleichungen für Summen, Reihen und Integrale, Funktionalgleichungen, Ungleichungen in der Approximationstheorie, in der Funktionsanalysis und in der Operatorentheorie, geometrisch und topologische Ungleichungen und Ungleichungen in der Wahrscheinlichkeits- und Informationstheorie. Ein sehr informativer Übersichtsartikel von P. L. Butzer (Aachen) berichtet über Erweiterungen des Banach-Steinhaus-Theorems, mit denen zusätzlich zur Konvergenz auch deren Geschwindigkeit erfaßt werden kann, und über zugehörige Anwendungen in verschiedenen Zweigen der Analysis. Die mathematischen Abhandlungen werden durch einen historischen Artikel von Frau Irmgard Süss über die Gründung und die erste Zeit des Mathematischen Forschungsinstituts in Oberwolfach aufgelockert. Die Autorin, Ehefrau von Wilhelm Süss, des Begründers des Forschungsinstituts, hat außerdem einige ihrer Zeichnungen zum Abdruck zur Verfügung gestellt, welche Ansichten vom Lorenzenhof, dem ursprünglichen Haus des Instituts, zeigen.

Dresden J. W. Schmidt

Newell, G. F., Approximate Behavior of Tandem Queues. Lecture Notes in Economics and Mathematical Systems 171. Berlin-Heidelberg-New York, Springer-Verlag 1979. XI, 410 S., 44 Abb., DM 39,—. US $ 21.50. ISBN 3-540-09552-7

Die Monographie will einen Beitrag zur Untersuchung von hintereinandergeschalteten Einbedienersystemen geben. Es wird versucht, Aussagen über die Struktur der Abhänigkeit der Charakteristiken des Tandemsystems von den Erwartungswerten und Varianzen der Bedienungszeiten sowie der Größe des Warteraumes der einzelnen Systeme zu gewinnen. Wichtigstes Hilfsmittel ist neben der deterministischen Approximation die Diffusionsapproximation. Durch die duale Betrachtung freier Warteplätze anstelle der Anzahl der Forderungen im System wird die Approximation auch ohne 'heavytraffic' -Annahmen interessant.

Die analytische und numerische Lösung der auftretenden Diffusionsprobleme nimmt den überwiegenden Teil der Arbeit in Anspruch, wobei nur der Fall höchstens zwei aufeinanderfolgender Systeme und nur spezieller Parameterkombinationen zu einer Lösung geführt werden konnte. Es wird sehr detailliert auf die Schwierigkeiten bei der Lösung sowie der Verallgemeinerung der Ergebnisse und Methoden eingegangen. Die zahlreichen graphischen Darstellungen der Resultate und die durchgeführten Grenzbetrachtungen sind hilfreich, insbesondere können sie für weitergehende qualitative Aussagen genutzt werden.

Freiberg U. Jansen

FSC
www.fsc.org
MIX
Papier | Fördert gute Waldnutzung
FSC® C083411

Zeitfracht Medien GmbH
Ferdinand-Jühlke-Straße 7
99095 Erfurt, Deutschland
produktsicherheit@kolibri360.de